T0211465

# LONDON MATHEMATICAL SOCIETY LECTURE NOTE SERIES

Managing Editor: Professor M. Reid, Mathematics Institute,
University of Warwick, Coventry CV4 7AL, United Kingdom

The titles below are available from booksellers, or from Cambridge University Press at www.cambridge.org/mathematics

London Mathematical Society Lecture Note Series: 360

# Zariski Geometries
## Geometry from the Logician's Point of View

**BORIS ZILBER**

*University of Oxford*

CAMBRIDGE
UNIVERSITY PRESS

# CAMBRIDGE
## UNIVERSITY PRESS

University Printing House, Cambridge CB2 8BS, United Kingdom

One Liberty Plaza, 20th Floor, New York, NY 10006, USA

477 Williamstown Road, Port Melbourne, VIC 3207, Australia

314-321, 3rd Floor, Plot 3, Splendor Forum, Jasola District Centre, New Delhi - 110025, India

79 Anson Road, #06-04/06, Singapore 079906

Cambridge University Press is part of the University of Cambridge.

It furthers the University's mission by disseminating knowledge in the pursuit of education, learning and research at the highest international levels of excellence.

www.cambridge.org
Information on this title: www.cambridge.org/9780521735605

First published 2010

*A catalogue record for this publication is available from the British Library*

*Library of Congress Cataloging in Publication data*
Zilber, Boris.
Zariski geometries : geometry from the logician's point of view / Boris Zilber.
p.   cm. – (London Mathematical Society lecture note series ; 360)
Includes bibliographical references and index.
ISBN 978-0-521-73560-5 (pbk.)
1. Zariski surfaces.   2. Geometry, Algebraic.   3. Zariski, Oscar, 1899–1986.
I. Title.   II. Series.
QA573.Z55   2010
516.3′5 – dc22
2009039896

ISBN 978-0-521-73560-5 Paperback

To Tamara, my wife

# Contents

# Acknowledgments

The work on this book started in 1991, and through all these years, many people helped me with their suggestions, questions, and critical remarks. First of all, in 1994 Kobi Peterzil used my very raw lecture notes to write a lecture course, which is now the core of Chapters 3 and 4 of this book. Practically all the exercises in these chapters are his work. Tristram de Piro read the lecture notes in 2000–1, and discussions with him and his further work on the topic had a big effect on the content of the book. A lot of the material in Chapter 6 is based on joint work with Nick Peatfield; explicit references are therein. Assaf Hasson and I worked on a problem related to the content of Chapter 6, and although the work did not result in a paper, it significantly contributed to the content of the chapter. This chapter has also been essentially influenced by the thesis written by Lucy Burton (Smith). Jonathan Kirby made many useful suggestions and remarks, particularly concerning Chapter 2. I am indebted to Matt Piatkus for the present form of Lemma 2.2.21. Kanat Kudajbergenov carefully read Chapters 2 and 3 of the book at an early stage and made many useful comments and corrections.

# 1

# Introduction

## 1.1 Introduction

The main purpose of writing this book is to convey to the general mathematical audience the notion of a Zariski geometry with the whole spectrum of geometric ideas arising in model-theoretic context. The idea of a Zariski geometry is intrinsically linked with algebraic geometry, as are many other model-theoretic geometric ideas. However, there are also very strong links with combinatorial geometries, such as matroids (pre-geometries) and abstract incidence systems. Model theory developed a very general unifying point of view based on the model-theoretic geometric analysis of mathematical structures as diverse as compact complex manifolds and general algebraic varieties, differential fields, difference fields, algebraic groups, and others. In all of these, Zariski geometries have been detected and have proved crucial for the corresponding theory and applications. In more recent works, this author has established a robust connection to non-commutative algebraic geometry.

Model theory has always been interested in studying the relationship between a mathematical structure, such as the field of complex numbers $(\mathbb{C}, +, \cdot)$, and its description in a formal language, such as the finitary language suggested by D. Hilbert: *the first-order language*. The best possible relationship is when a structure $M$ is the unique, up-to-isomorphism model of the description $\text{Th}(M)$: *the theory of $M$*. Unfortunately, for a first-order language, this is the case only when $M$ is finite because in the first order, it is impossible to fix an infinite cardinality of (the universe of) $M$. So, the next best relationship is when the isomorphism type of $M$ is determined by $\text{Th}(M)$ and the cardinality $\lambda$ of $M$ ($\lambda$-categoricity), such as $\text{Th}(\mathbb{C}, +, \cdot)$ – the theory of the field of complex numbers, in which 'complex algebraic geometry lives'. Especially interesting is the case when $\lambda$ is uncountable and the description is at most countable. In fact, in this case, Morley's theorem (1965) states that the theory Th($M$) is not sensitive to a particular choice of $\lambda$; it has a unique model in every uncountable cardinality.

The proof of Morley's theorem marked the beginning of stability theory, which studies categorical theories in uncountable cardinals and generalisations (*every categorical theory in uncountable cardinals is stable*). Categoricity and stability turned out to be amazingly effective classification principles. To sum up the results of 40 years of research in a few lines, we lay out the following conclusions:

1. There is a clear hierarchy of the 'logical perfection' of a theory in terms of stability. Categorical theories and their models are at the top of this hierarchy.
2. The key feature of stability theory is dimension theory and, linked to it, dependence theory resembling the dimension theory of algebraic geometry and the theory of algebraic dependence in fields. In fact, algebraic geometry and its related areas are the main sources of examples.
3. There has been considerable progress toward the classification of structures with stable and, especially, uncountably categorical theories. The (fine) classification theory makes use of certain geometric principles, both classical and those specifically developed in model theory. These geometric principles proved useful in applications, such as in Diophantine geometry.

In classical mathematics, three basic types of dependencies are known:

1. algebraic dependence in the theory of fields;
2. linear dependence in the theory of vector spaces; and
3. dependence of trivial (combinatorial) type (e.g. two vertices of a graph are dependent if they belong to the same connected component).

One of the useful conjectures in fine classification theory was the **trichotomy principle**, which states that every dependence in an uncountably categorical theory is based on one of three classical types.

A more elaborate form of this conjecture implies that any uncountably categorical structure with a non-linear, non-trivial geometry comes from algebraic geometry over an algebraically closed field. (It makes sense to call a dependence type *non-linear* if it does not belong to types 2 and 3.) For example, a special case of this conjecture has been known since 1975 and is still open (see survey by Cherlin, 2002).

**The algebraicity conjecture**: Suppose $(G, \cdot)$ is a simple group with Th($G$) categorical in uncountable cardinals. Then $G = \mathrm{G}(K)$ for some simple algebraic group G and an algebraically closed field $K$.

The trichotomy principle proved to be false in general (Hrushovski, 1988) but nevertheless holds for many important classes. The notion of a Zariski structure was designed primarily to identify all such classes.

Originally, the idea of a Zariski structure was a condition which would isolate the 'best' possible classes on the top of the hierarchy of stable structures. Because it has been realised that purely logical conditions are not sufficient for the trichotomy principle to hold, it has also been realised that a topological ingredient added to the definition of a categorical theory might suffice. In fact, a very coarse topology similar to the Zariski topology in algebraic geometry is sufficient. Along with the introduction of the topology, one also postulates certain properties of it, mainly of how the topology interacts with the dimension notion. One of the crucial properties of this kind is in fact a weak form of smoothness of the geometry in question; in this book, it is called *the pre-smoothness property.*

In more detail, a (Noetherian) Zariski structure is a structure $M = (M, \mathcal{C})$, on the universe $M$ in the language given by the family of relations listed in $\mathcal{C}$.

For each $n$, the subsets of $M^n$ corresponding to relations from $\mathcal{C}$ form a Noetherian topology. The topology is endowed with a dimension notion (e.g. the Krull dimension). Dimension is well behaved with respect to projections $M^{n+1} \rightarrow M^n$.

The structure M is said to be pre-smooth if for any two closed irreducible $S_1, S_2 \subseteq M^n$, and for any irreducible component $S_0$ of the set $S_1 \cap S_2$,

$$\dim S_0 \geq \dim S_1 + \dim S_2 - \dim M^n.$$

It has been said already that the basic examples of pre-smooth Noetherian Zariski structures come from algebraic geometry. Indeed, let $M = M(K)$ be the set of $K$-points of a smooth algebraic variety over an algebraically closed field $K$. For $\mathcal{C}$, take the family of Zariski closed subsets (relations) of $M^n$, all $n$. Set dim $S$ to be the Krull dimension. This is a pre-smooth Zariski structure (geometry).

Another important class of examples is the class of compact complex manifolds. Here $M$ should be taken to be the underlying set of a manifold, and $\mathcal{C}$ should be taken as the family of all analytic subsets of $M^n$, all $n$.

Proper analytic varieties in the sense of *rigid analytic geometry* (analogues of compact complex manifolds for non-Archimedean valued fields) constitute yet another class of Noetherian Zariski structures.

It follows from the general theory developed in these lectures that all these structures (and Zariski structures in general) are at the top of the logical hierarchy (i.e. they have finite Morley rank and in most important cases are uncountably categorical). Interestingly, for the second and third classes, this is hard to establish without first checking that the structures are Zariski.

So far, the main result of the general theory is the classification of one-dimensional, pre-smooth, Noetherian Zariski geometries M:

If M *is non-linear, then there is an algebraically closed field* $K$, *a quasi-projective algebraic curve* $C_M = C_M(K)$, *and a surjective map*

$$p : M \to C_M$$

*of a finite degree (i.e.* $p^{-1}(a) \leq d$ *for each* $a \in C_M$) *such that for every closed* $S \subseteq M^n$, *the image* $p(S)$ *is Zariski closed in* $C_M^n$ *(in the sense of algebraic geometry); if* $\hat{S} \subseteq C_M^n$ *is Zariski closed, then* $p^{-1}(\hat{S})$ *is a closed subset of* $M^n$ *(in the sense of the Zariski structure M).*

In other words, M is almost an algebraic curve. In fact, it is possible to specify some extra geometric conditions for M which imply M is an exact algebraic curve (see [HZ]).

The proof of the classification theorem proceeds as follows (Chapters 3 and 4):

First, for general Zariski structures, we develop an *infinitesimal analysis* that culminates with the introduction of *local multiplicities* of covers (maps) and intersections and the proof of *the implicit function theorem*.

Next, we focus on a specific configuration in a one-dimensional M given by the two-dimensional pre-smooth 'plane' $M^2$ and an $n$-dimensional ($n \geq 2$) pre-smooth family $L$ of curves on $M^2$. We use the local multiplicities of intersections to define what it means to say that two curves are *tangent at a given point*. This is well defined in non-singular points of the curves, but in general we need a more subtle notion. This is a technically involved concept of a *branch of a curve at a point*. Once this is properly defined, we develop a theory of tangency for branches and prove, in particular, that tangency between branches is an equivalence relationship.

Next, we treat branches of curves on the plane $M^2$ as (graphs of) *local functions* from an infinitesimal neighbourhood of a point on $M$ onto another infinitesimal neighbourhood. One can prove that the composition of such local functions is well behaved with respect to tangency. In particular, with respect to composition modulo tangency, local functions form a local group (pre-group, or a 'group-chunk' in the terminology of A. Weil). A generalisation of a known proof by Weil produces a pre-smooth *Zariski group*, more specifically an Abelian group $J$ of dimension 1.

We now replace the initial one-dimensional $M$ by the more suitable Zariski curve $J$ and repeat the construction on the plane $J^2$. Again, we consider the composition of local functions on $J$ modulo tangency. However, this time we take into account the existing group structure on $J$ and find that our new group operation interacts with the existing one in a nice way. More specifically, the new group structure acts (locally) on the existing one by (local) endomorphisms. Using again the generalisation of Weil's pre-group theorem, we find *a field K with a Zariski structure on it.*

Notice that at this stage we do not know if the Zariski structure on $K$ is the classical (algebraic) one. Obviously, it contains all algebraic Zariski closed relations, but we need to see that there are no extra ones in the Zariski topology. For this purpose, we undertake an analysis of projective spaces $\mathbf{P}^n(K)$. First, we prove that $\mathbf{P}^n(K)$ are *weakly complete* in our Zariski topology, which is the property analogous to the classical completeness in algebraic geometry. Then, expanding the intersection theory of the first sections, we manage to prove a *generalisation of the Bezout theorem*. This theorem is key in proving *the generalisation of the Chow theorem*: every Zariski closed subset of $\mathbf{P}^n(K)$ is algebraic. (Note that $\mathbf{P}^n(\mathbb{C})$ is a compact complex manifold and that every analytic subset of it is Zariski closed according to our definition.) This immediately implies that the structure on $K$ is purely algebraic.

It follows from the construction of $K$ in M that there is a non-constant Zariski-continuous map $f : M \to K$, with the domain of definition open in $M$. Such maps we call *Z-meromorphic functions*. Based on the generalisation of Chow's theorem, we prove that the inseparable closure of the field $K_Z(M)$ of Z-meromorphic functions is isomorphic to the field of rational functions of a smooth algebraic curve $C_M$. By the same construction, we find a Zariski-continuous map $p : M \to C_M$ which satisfies the required properties. This completes the proof of the classification theorem.

The classification theorem asserts that in the one-dimensional case, a nonlinear Zariski geometry is *almost* an algebraic curve. This statement is true completely in algebraic geometry, compact complex manifolds, and proper rigid analytic varieties; in the last two, this is due to the Riemann existence theorem. However, in the general context of Noetherian Zariski geometries, the adverb 'almost' cannot be omitted. In Section 5.1, we present a construction that provides examples of non-classical Noetherian Zariski geometries, that is, ones which are *not definable in an algebraically closed field*. We study a special but typical example and look for a way to explain the geometry of M in terms of co-ordinate functions to $K$ and co-ordinate rings. We conclude that there are just not enough of regular (definable Zariski-continuous) functions $M \to K$ and that we need to use a larger class of functions, *semi-definable coordinate functions* $\phi : M \to K$. We introduce a $K$-vector space $\mathcal{H}$ generated by these functions and define linear operators on $\mathcal{H}$ corresponding to the actions by $\tilde{G}$. These generate a non-commutative $K$-algebra $A$ on $\mathcal{H}$. Importantly, $A$ is determined uniquely (up to the choice of the language) in spite of the fact that $\mathcal{H}$ is not. Also, a non-trivial, semi-definable function induces on $K$ some extra structure, which we call here $*$-data. Correspondingly, this adds some extra structure to the $K$-algebra $A$, which eventually makes it a $C^*$-algebra. Finally, we are able to recover the M from $A$. Namely, M is identified with the set of eigenspaces of 'self-adjoint' operators of $A$ with the Zariski topology given by certain ideals of $A$. In other words, this new and more general class of Zariski geometries can be appropriately explained in terms of non-commutative co-ordinate rings.

We then discuss further links to non-commutative geometry. We show how, given a typical *quantum algebra A at roots of unity*, one can associate a Zariski geometry with $A$. This is similar to, although slightly different from, the connection between M and $A$ in the preceding discussion. Importantly, for a typical non-commutative $A$, the geometry turns out to be non-classical, whereas for a commutative one, it is equivalent to the classical affine variety Max $A$.

The final chapter introduces a generalisation of the notion of a Zariski structure. We call the more general structures *analytic Zariski*. The main difference is that we no longer assume the Noetherianity of the topology. This makes the definition more complicated because we now have to distinguish between general closed subsets of $M^n$ and the ones with better properties, which we call analytic. The main reward for the generalisation is that now we have a much wider class of classical structures (e.g. universal covers of some algebraic varieties) that satisfy the definition. One hope (which has not been realised so far) is to find a way to associate an analytic Zariski geometry with a generic quantum algebra.

The theory of analytic Zariski geometries is still in its infancy. We do not know if the algebraicity conjecture is true for analytic Zariski groups, which is an interesting and important problem. One of the main results presented here is the theorem stating that any compact, analytic Zariski structure is Noetherian, that is, it satisfies the basic definition. We also prove some model-theoretic properties of analytic Zariski structures, establishing their high level in the logical hierarchy, but remarkably this is the non-elementary logic stability hierarchy formulated in terms of Shelah's *abstract elementary classes*. This is a relatively new domain of model theory, and analytic Zariski structures constitute a large class of examples for this theory.

We hope that these notes may be useful not only for model theorists but also for people who have a more classical, geometric background. For this reason, we start the notes with a crash course in model theory. It is really basic, and the most important thing to learn in this section is the spirit of model theory. The emphasis on the study of definability with respect to a formal language is perhaps central to doing mathematics in a model-theoretic way.

## 1.2 About model theory

This section gives a very basic overview of model-theoretic notions and methods. We hope that the reader will be able to grasp the main ideas and the spirit of the subject. We did not aim to give proofs in this section of every statement we found useful, and even definitions are missing some detail. To compensate for this, in Appendix A we give a detailed list of basic model-theoretic facts, definitions, and proofs. Appendix B surveys geometric stability theory and some more recent results relevant to the material in the main chapters.

Of course, there is a good selection of textbooks on model theory. The most adequate for our purposes is Marker (2002), and a more universal book is Hodges (1993).

The crucial feature of the model-theoretic approach to mathematics is the attention paid to the formalism with which one considers particular mathematical structures.

A structure M is given by a set $M$, the **universe** (or the **domain**) of M, and a family $L$ of relations on $M$, called **primitives of $L$** or **basic relations**. One often writes $M = (M, L)$. $L$ is called the **language** for M.

Each relation has a fixed name and arity, which allows us to consider classes of $L$-structures of the form $(N, L)$, where $N$ is a universe and $L$ is the collection of relations on $N$ with the names and arities fixed (by $L$). Each such structure $(N, L)$ represents an **interpretation** of the language $L$.

Recall that an $n$-ary relation $S$ on $M$ can be identified with a subset $S \subseteq M^n$. When $S$ is just a singleton $\{s\}$, the name for $S$ is often called a **constant symbol** of the language. One can also express functions in terms of relations; instead of saying $f(x_1, \ldots, x_n) = y$, one says just that $\langle x_1, \ldots, x_n, y \rangle$ satisfies the $(n + 1)$-ary relation $f(x_1, \ldots, x_n) = y$. There is no need to include special function and constant symbols in $L$.

One always assumes that the binary relation $=$ is in the language and is interpreted canonically.

**Definition 1.2.1.** The following is an inductive definition of a **definable set** in an $L$-structure M:

(i)  a set $S \subseteq M^n$ interpreting a primitive $S$ of the language $L$ is definable;
(ii)  given definable $S_1 \subseteq M^n$ and $S_2 \subseteq M^m$, the set $S_1 \times S_2 \subseteq M^{n+m}$ is definable (here $S_1 \times S_2 = \{x^\frown y : x \in S_1, \, y \in S_2\}$);

(iii) given definable $S_1, S_2 \subseteq M^n$, the sets $S_1 \cap S_2$, $S_1 \cup S_2$, and $M^n \setminus S_1$ are definable; and

(iv) given definable $S \subseteq M^n$ and a projection $\text{pr} : \langle x_1, \ldots, x_n \rangle \mapsto \langle x_{i_1}, \ldots, x_{i_m} \rangle$, $\text{pr} : M^n \to M^m$, the image $\text{pr}\, S \subseteq M^m$ is definable.

Note that item (iv), for $n = m$, allows a permutation of variables.

The definition can also be applied to definable functions, definable relations, and even definable points.

An alternative but equivalent definition is given by introducing the (first-order) $L$-formulas. In this approach, we write $S(x_1, \ldots, x_n)$ instead of $\langle x_1, \ldots, x_n \rangle \in S$, starting from basic relations, and then we construct arbitrary formulas by induction using the logical connectives $\wedge, \vee$, and $\neg$ and the quantifier $\exists$.

Now, given an $L$-formula $\psi$ with $n$ free variables, the set of the form

$$\psi(M^n) := \{\langle x_1, \ldots, x_n \rangle \in M^n : \text{M} \vDash \psi(x_1, \ldots, x_n)\},$$

is said to be definable (by formula $\psi$).

The approach via formulas is more flexible because we may use formulas to define sets with the same formal description, say $\psi(N^n)$, in arbitrary $L$-structures.

Moreover, if formula $\psi$ has no free variables (then called a **sentence**), it describes a property of the structure itself. In this way, classes of $L$-structures can be defined by axioms in the form of $L$-formulas.

One says that N is **elementarily equivalent** to M (written $\text{N} \equiv \text{M}$) if for all $L$-sentences $\varphi$

$$\text{M} \vDash \varphi \Leftrightarrow \text{N} \vDash \varphi.$$

**Example 1.2.2.** Groups can be considered $L$-structures where $L$ has one constant symbol $e$ and one ternary relation symbol $P(x, y, z)$, interpreted as $x \cdot y = z$. For example, the associativity property then can be written as

$$\forall x, y, z, u, v, w, t \ (P(x, y, u) \wedge P(u, z, v) \wedge P(x, w, t) \wedge P(y, z, w) \rightarrow v = t).$$

Here $\forall x\, A$ means $\neg \exists x\, \neg A$, and the meaning of $B \rightarrow C$ is $\neg B \vee C$.

The centre of a group $G$ can be defined as $\varphi(G)$, where $\varphi(x)$ is the formula

$$\forall y, z \ (P(x, y, z) \leftrightarrow P(y, x, z)).$$

Of course, this definable set can be described in line with Definition 1.2.1, although this would be slightly longer description.

One important advantage of Definition 1.2.1 is that it provides a more geometric description of the set. We use both approaches interchangeably.

One of the most useful types of model-theoretic results is a **quantifier elimination** statement. One says that M (or more usually, the theory of M) has quantifier elimination if any definable set $S \subseteq M^n$ is of the form $S = \psi(M^n)$ where $\psi(\bar{x})$ is a **quantifier-free** formula, that is, one obtained from primitives of the language using connectives but no quantifiers.

**Example 1.2.3.** Define the language $L_{Zar}$ with primitives given by zero-sets of polynomials over the prime subfield.

**Theorem (Tarski, also Seidenberg and Chevalley).** *An algebraically closed field has quantifier elimination in language $L_{Zar}$.*

Recall that in algebraic geometry, a Boolean combination of zero-sets of polynomials (Zariski closed sets) is called a constructible set. The theorem says, in other words, that the class of definable sets in an algebraically closed field is the same as the class of constructible sets.

Note that for each $S$, the fact that $S = \psi(M^n)$ is expressible by the $L$-sentence $\forall \bar{x}(S(\bar{x}) \leftrightarrow \psi(\bar{x}))$. Hence, quantifier elimination holds in M if and only if it holds in any structure elementarily equivalent to M.

Given a class of elementarily equivalent $L$-structures, the adequate notion of embedding is that of an **elementary embedding**. We say that $M = (M, L)$ is an elementary substructure of $M' = (M', L)$ if $M \subseteq M'$ and for any $L$-formula $\psi(\bar{x})$ with free variables $\bar{x} = \langle x_1, \ldots, x_n \rangle$ and any $\bar{a} \in M^n$,

$$M \vDash \psi(\bar{a}) \Leftrightarrow M' \vDash \psi(\bar{a}).$$

More generally, elementary embedding of M into $M'$ means that M is isomorphic to an elementary substructure of $M'$. We write the elementary embedding (elementary extension) as

$$M \preccurlyeq M'.$$

Note that $M \preccurlyeq M'$ always implies that $M \equiv M'$, because an elementary embedding preserves all $L$-formulas, including sentences.

**Example 1.2.4.** Let $\mathbb{Z}$ be the additive group of integers in the group language of Example 1.2.2. Obviously $z \mapsto 2z$ embeds $\mathbb{Z}$ into itself as $2\mathbb{Z}$. However, this is not an elementary embedding because the formula $\exists y \, y + y = x$ holds for $x = 2$ in $\mathbb{Z}$ but does not hold in the substructure $2\mathbb{Z}$. On the other hand, for

$K \subseteq K'$ algebraically closed fields in language $L_{Zar}$, the embedding is always elementary. This is an immediate result of the quantifier elimination theorem.

A simple but useful technical fact is given by the following.

**Exercise 1.2.5.** *Let*

$$M_1 \prec \cdots M_\alpha \prec M_{\alpha+1} \prec \cdots$$

*be an ascending sequence of elementary extensions, $\alpha \in I$, and let*

$$^*M = \bigcup_{\alpha \in I} M_\alpha$$

*be the union. Then, for each $\alpha \in I$, $M_\alpha \prec {}^*M$.*

When we want to specify an element in a structure M in terms of $L$, we describe its **type**. Given $\bar{a} \in M^n$, the **type** of $\bar{a}$ is the set of $L$-formulas with $n$ free variables $\bar{x}$:

$$\mathrm{tp}(\bar{a}) = \{\psi(\bar{x}) : \ M \vDash \psi(\bar{a})\}.$$

Often we look for $n$-tuples, in M or its elementary extensions, that satisfy a certain description in terms of $L$. For this purpose, one uses a more general notion of a type.

**Definition 1.2.6.** An $n$-type in M is a set $p$ of $L$-formulas $\psi(\bar{x})$ (with free variables $\bar{x} = \langle x_1, \ldots, x_n \rangle$) satisfying the consistency condition:

$$\psi_1(\bar{x}), \ldots, \psi_k(\bar{x}) \in p \Rightarrow M \vDash \exists \bar{x}\, \psi_1(\bar{x}) \wedge \cdots \wedge \psi_k(\bar{x}).$$

Obviously, the M in the consistency condition can be equivalently replaced by any M′ elementarily equivalent to M.

**Example 1.2.7.** Let $\mathbb{R}$ be the field of reals in language $L_{Zar}$. Note that the relation $x \leq y$ is expressible in $\mathbb{R}$ by the formula $\exists u\, u^2 + x = y$. So, in the language we can write down the type of a real positive **infinitesimal**,

$$p = \left\{ 0 < x < \frac{1}{n} : \ n \in \mathbb{Z}, n > 0 \right\}.$$

Obviously, this type is not realised in $\mathbb{R}$ itself, but there is $\mathbb{R}' \succ \mathbb{R}$, which realises $p$.

Often, we have to consider $L$-formulas **with parameters**. For example, in Example 1.2.3, the basic relations are given by polynomial equations over the prime field, but one usually is interested in polynomial equations over $K$. Clearly, this can be achieved within the same language if we use parameters: if $P(x_1, \ldots, x_n, y_1, \ldots, y_m)$ is a polynomial equation over the prime field and

$a_1, \ldots, a_m \in K$, then $P(x_1, \ldots, x_n, a_1, \ldots, a_m)$ is a polynomial equation over $K$, and all polynomial equations over $K$ can be obtained in this way. There is no need to develop an extra theory to deal with formulas with parameters; these can be seen simply as formulas of the extended language $L(C)$, where $C$ is the set of parameters we want to use. For example, if we want to use formulas with parameters in $M$, the universe of M, the language will be called $L(M)$, and it will consist of the primitives of $L$ plus one singleton primitive (constant symbol) for each element of $M$.

Formulas and types of language $L(C)$ are also called **formulas and types over** $C$. For corresponding sets, we often say only **$C$-definable**. When this terminology is used, 0-definable means definable without parameters.

A basic but very useful theorem of model theory is the **compactness theorem** (A. Mal'tsev, 1936). In its basic form, it states that *any finitely satisfiable set of L-sentences has a model.* Here, finitely satisfiable means that any finite subset of the set of sentences has a model.

In this book, we usually work with a given structure M and its elementary extensions. More useful in this situation is the following corollary of the compactness theorem:

**Theorem 1.2.8 (corollary of the compactness theorem).** *Let* M *be an L-structure and* P *a set of types over* M. *Then there is an elementary extension* $M' \succeq M$ *in which all the types of P are realised. Moreover, we can choose* M' *to be of cardinality not bigger than* $\max\{cardM, cardP\}$.

In particular, we can choose $P$ to be the set of all $n$-types over M, all $n$. Then, any M$'$ realising $P$ will be said to be **saturated over** M.

Consider

$$M_0 \preccurlyeq M_1 \preccurlyeq \cdots M_i \preccurlyeq \cdots$$

to be an ascending chain of elementary extensions, $i \in \mathbb{N}$, such that $M_{i+1}$ realises all the types over finite subsets of $M_i$. Then, $\bigcup_i M_i$ has the property that *every n-type over a finite subset of the structure is realised in this structure.* A structure with this property is said to be $\aleph_0$**-saturated** (or $\omega$-saturated).

More generally, a structure M in which every $n$-type over a subset of cardinality less than $\kappa$ is realised in the structure is said to be $\kappa$**-saturated**.

It is easy to see that all these definitions remain equivalent if we ask just for 1-types to be realised.

# 2

# Topological structures

## 2.1 Basic notions

Let M be a structure and let $\mathcal{C}$ be a distinguished sub-collection of the definable subsets of $M^n$, $n = 1, 2, \ldots$ The sets in $\mathcal{C}$ are called (definable) **closed**. The relations corresponding to the sets are the basic (primitive) relations of the language we work with. $\langle M, \mathcal{C} \rangle$, or M, is a **topological structure** if it satisfies the following axioms.

(L) **Language:** The primitive $n$-ary relations of the language are exactly the ones that distinguish definable closed subsets of $M^n$, all $n$ (that is, the ones in $\mathcal{C}$), and every quantifier-free positive formula in the language defines a closed set. More precisely;

1. the intersection of any family of closed sets is closed;
2. finite unions of closed sets are closed;
3. the domain of the structure is closed;
4. the graph of equality is closed;
5. any singleton of the domain is closed;
6. Cartesian products of closed sets are closed;
7. the image of a closed $S \subseteq M^n$ under a permutation of co-ordinates is closed; and
8. for $a \in M^k$ and $S$, a closed subset of $M^{k+l}$ defined by a predicate $S(x, y)$ ($x = \langle x_1, \ldots, x_k \rangle$, $y = \langle y_1, \ldots, y_l \rangle$), the set $S(a, M^l)$ (the *fibre over a*) is closed.

Here and in what follows the fibre over $a$

$$S(a, M^l) = \{b \in M^l : M \models S(a, b)\}$$

12

and *projections* are the maps

$$\mathrm{pr}_{i_1,\ldots,i_m} : \langle x_1, \ldots, x_n \rangle \mapsto \langle x_{i_1}, \ldots, x_{i_m} \rangle, \quad i_1, \ldots, i_m \in \{1, \ldots, n\}.$$

L6 needs some clarification. If $S_1 \subseteq M^n$ and $S_2 \subseteq M^m$ are closed, the assumption states that $S_1 \times S_2$ canonically identified with a subset of $M^{n+m}$ is closed in the latter. The canonical identification is

$$\langle \langle x_1, \ldots, x_k \rangle, \langle y_1, \ldots, y_m \rangle \rangle \mapsto \langle x_1, \ldots, x_k, y_1, \ldots, y_m \rangle.$$

**Remark 2.1.1.** A projection $\mathrm{pr}_{i_1,\ldots,i_m}$ is a continuous map in the sense that the inverse image of a closed set $S$ is closed. Indeed,

$$\mathrm{pr}_{i_1,\ldots,i_m}^{-1} S = S \times M^{n-m}$$

up to the order of co-ordinates.

**Exercise 2.1.2.** *Prove that, given $a \in M^k$, the bijection*

$$\mathrm{concat}_a^n : x \in M^n \mapsto a^\frown x \in (\{a\} \times M^n)$$

*is a homeomorphism $M^n \to \{a\} \times M^n$; that is, the closed subsets on $(M^n)^m$ correspond to closed subsets on $(\{a\} \times M^n)^m$ and conversely for all $m$.*

We sometimes refer to definable subsets of $M^n$ as logical predicates. For example, we may say $F(a)$ instead of saying $a \in F$, or $S_1(x) \ \& \ S_2(x)$ instead of $S_1 \cap S_2$.

We write $U \subseteq_{\mathrm{op}} M^n$ to say that $U$ is open in $M^n$ and $S \subseteq_{\mathrm{cl}} M^n$ to say that it is closed.

**Constructible sets** are by definition the Boolean combinations of members of $\mathcal{C}$. It is easy to see that a constructible subset of $M^n$ can be equivalently described as a finite union of sets $S_i$, such that $S_i \subseteq_{\mathrm{cl}} U_i \subseteq_{\mathrm{op}} M^n$.

A subset of $M^n$ is called **projective** if it is a finite union of sets of the form $\mathrm{pr}\, S_i$, for some $S_i \subseteq_{\mathrm{cl}} U_i \subseteq_{\mathrm{op}} M^{n+k_i}$ and projections $\mathrm{pr}^{(i)} : M^{n+k_i} \to M^n$.

Note that any constructible set is projective with trivial projections in its definition.

A topological structure is said to be **complete** if (P) **properness** of projections condition holds: the image $\mathrm{pr}_{i_1,\ldots,i_m}(S)$ of a closed subset $S \subseteq_{\mathrm{cl}} M^n$ is closed.

A topological structure $M$ will be called **quasi-compact** (or just **compact**) if it is complete and satisfies (QC): for any finitely consistent family $\{C_t : t \in T\}$ of closed subsets of $M^n$

$$\bigcap_{t \in T} C_t$$

is non-empty.

The same terminology can be used for a subset $S \subseteq M^n$ if the induced topological structure satisfies (P) and (QC).

Notice that (QC) is equivalent to saying that every open cover of $M^n$ has a finite sub-cover.

A topological structure is called **Noetherian** if it also satisfies (DCC): **descending chain condition** for closed subsets. For any closed

$$S_1 \supseteq S_2 \supseteq \ldots S_i \supseteq \ldots,$$

there is $i$ such that for all $j \geq i$, $S_j = S_i$.

A definable set $S$ is called **irreducible** if there are no relatively closed subsets $S_1 \subseteq_{cl} S$ and $S_2 \subseteq_{cl} S$ such that $S_1 \subsetneq S_2$, $S_2 \subsetneq S_1$, and $S = S_1 \cup S_2$.

**Exercise 2.1.3.** *(DCC) implies that for any closed $S$ there are distinct closed irreducible $S_1, \ldots, S_k$ such that*

$$S = S_1 \cup \cdots \cup S_k.$$

*These $S_i$ are called **irreducible components of** $S$. They are defined uniquely up to numeration.*

We can also consider a decomposition $S = S_1 \cup S_2$ for $S$ constructible and $S_1, S_2$ closed in $S$. If there is no proper decomposition of a constructible $S$, we say that $S$ is irreducible.

**Exercise 2.1.4.** *Let* pr $: M^n \to M^m$ *be a projection.*

*(i) For $S \subseteq_{cl} M^n$, pr $: M^n \to M^k$, $\overline{\text{pr} \, S}$ is irreducible if $S$ is irreducible.*

*(ii) If M is Noetherian and $\overline{\text{pr} \, S}$ is irreducible, then $\overline{\text{pr} \, S} = \overline{\text{pr} \, S'}$ for some irreducible component $S'$ of S.*

**Exercise 2.1.5.**

1. *For every $n$ and $k$, the topology on $M^{n+k}$ extends the product topology on $M^n \times M^k$.*
2. *If $S_1$, $S_2$ are closed irreducible sets, then pr $S_1$ and $S_1 \times S_2$ are irreducible as well.*
3. *If $F(x, y)$ is a relation defining a closed set, then $\forall y F(x, y)$ defines a closed set as well.*

## 2.2 Specialisations

We introduce here one of the main tools of the theory, which we call a *specialisation*. It has analogues both in model theory and algebraic geometry. In the latter, the notion under the same name has been used by A. Weil (1967); namely, if $K$ is an algebraically closed field and $\bar{a}$ is a tuple in an extension $K'$ of $K$, then a mapping $K[\bar{a}] \to K$ is called a specialisation if it preserves all equations with coefficients in $K$.

In the same setting, a specialisation is often called a *place*.

The model-theoretic source of the notion is A. Robinson's *standard-part map* from an elementary extension of $\mathbb{R}$ (or $\mathbb{C}$) onto the compactification of the structure; see Example 2.2.4. The concept emerges in model theory in the context of *atomic compact* structures, introduced by J. Mycielski (1964) and given a thorough study by B. Weglorz (1966) and others.

A structure $M$ is said to be **atomic (positive) compact** if any finitely consistent set of atomic (positive) formulas is realized in $M$. It can be easily seen that Noetherian topological structures are atomic compact. The main result of the work of Weglorz (1966) (see also Hodges, 1993, for a proof) is the following theorem.

**Theorem 2.2.1 (B. Weglorz).** *The following are equivalent for any structure* M:

  (i) M *is atomic compact;*
 (ii) M *is positive compact; and*
(iii) M *is a retract of any* M' $\succ$ M; *that is, there is a surjective homomorphism* $\pi : M' \to M$, *fixing* M *pointwise,*

Our proofs of Propositions 2.2.7 and 2.2.10 are essentially based on the Weglorz's arguments.

Throughout this section, we assume that $M$ is just a topological structure.

**Definition 2.2.2.** Let $^*M \succeq M$ be an elementary extension of M and $M \subseteq A \subseteq {}^*M$. A map $\pi : A \to M$ is called a **(partial) specialisation** if for every $a$ from $A$ and an $n$-ary $M$-closed $S$, $a \in {}^*S$ and then $\pi(a) \in S$, where $^*S$ stands for the set of realisations of the relation $S$ in $^*M$, equivalently $S(^*M)$.

**Remark 2.2.3.** By definition, a specialisation is an identity on $M$, because any singleton $\{s\}$ is closed.

**Example 2.2.4.** Take $^*M = \langle {}^*\mathbb{R}, +, \cdot \rangle$, a non-standard elementary extension of $\langle \mathbb{R}, +, \cdot \rangle$. $^*M$ interprets in the obvious way an elementary extension of $\langle \mathbb{C}, +, \cdot \rangle$, where the universe of the latter is $\mathbb{R}^2$. Let $\tilde{\mathbb{R}} \subseteq {}^*M$ be the convex hull of $\mathbb{R}$ with respect to the linear ordering. There is then a map $st : \tilde{\mathbb{R}} \to \mathbb{R}$, which is known as the **standard-part map**. It is easy to see that $st$ induces a total map from $^*\mathbb{R} \cup \{\infty\} = \mathbf{P}^1(^*\mathbb{R})$ onto $\mathbb{R} \cup \{\infty\} = \mathbf{P}^1(\mathbb{R})$. In the same way, *st* induces a total map

$$\pi : \mathbf{P}^1(^*\mathbb{C}) \to \mathbf{P}^1(\mathbb{C}).$$

Using standard arguments, it is not difficult to see that $\pi$ is indeed a specialisation of Zariski structures.

**Example 2.2.5.** Let $K$ be an algebraically closed field and $^*K = K\{t^{\mathbb{Q}}\}$ the field of Puiseux series over $K$, the formal expressions

$$f = \sum_{n \geq m}^{\infty} a_n t^{\frac{n}{k}},$$

with $a_n \in K$, $n \in \mathbb{Z}$, and $k \in \mathbb{N}$ are fixed for each sum. This is known to be an algebraically closed field, so $K \prec {}^*K$.

Assuming $a_m \neq 0$, let $v(f) = m/k$. Then, $v : {}^*K \to \mathbb{Q}$ is a valuation. The valuation ring $R_v$ is equal to the ring of power series with non-negative valuation; that is, $m \geq 0$. For $f \in R_v$, set the coefficient $a_0 = a_0(f)$ to be 0 in the case $m > 0$. Then the map $\pi : R_v \to K$ defined as $\pi(f) := a_0(f)$ is a specialisation. One can extend it to the total specialisation $\mathbf{P}^1(^*K) \to \mathbf{P}^1(K)$ by setting $\pi(f) = \infty$ when $f \notin R_v$.

**Example 2.2.6.** The same is true if we replace $K\{t^{\mathbb{Q}}\}$ by $K\{t^{\Gamma}\}$ where $\Gamma$ is an ordered divisible Abelian group and the support of power series is a well-ordered field of generalised power series.

**Proposition 2.2.7.** *Suppose* M *is a quasi-compact structure,* $^*M \succeq M$. *Then, there is a total specialisation* $\pi : {}^*M \to M$. *Moreover, any partial specialisation can be extended to a total one.*

*Proof.* Define $\pi$ on $M \subseteq {}^*M$ as the identity map. Suppose $\pi$ is defined on some $D \subseteq {}^*M$ and $b'$ an element of $^*M$. We want to extend the definition of $\pi$ to $D \cup \{b'\}$. To do this, consider the set of formulas

$$\{S(x, d) : \ S \text{ is a closed relation, } {}^*M \models S(b', d'), \ d' \in D^n, \ d = \pi(d')\}.$$

Each $S(M, d)$ is non-empty. Indeed, $d'$ is in $\exists x S(x, {}^*M)$, which is closed by (P), so $M \models \exists x S(x, \pi(d'))$. Similarly, every finite intersection of such sets is non-empty. Thus, by (QC), the intersection of the sets in the family is non-empty. Put $\pi(b') = b$ for a $b$ chosen in the intersection. Immediately the extended map is again a partial specialisation. By continuing this process, one gets a total specialisation. $\square$

**Definition 2.2.8.** Given a (partial) specialisation, $\pi : {}^*M \to M$, we call a definable set (relation) $P \subseteq M^n$ $\pi$-**closed** if $\pi({}^*P) \subseteq P$. (In fact, $\subseteq$ can be replaced here by $=$.)

The family of all $\pi$-closed relations is denoted $\mathcal{C}_\pi$.

**Exercise 2.2.9.** $\mathcal{C}_\pi$ *satisfies (L); thus, M becomes a $\mathcal{C}_\pi$-structure.*

**Proposition 2.2.10.** *Given a total specialisation $\pi : {}^*M \to M$, the $\mathcal{C}_\pi$-structure M is quasi-compact provided that ${}^*M$ is saturated over M.*

*Proof.* The completeness (P) of the topology is immediate, because any map commutes with the projection.

To check (QC), consider any finitely consistent family $\{C_t : t \in T\}$ of closed subsets of $M^n$. Then $\bigcap\{{}^*C_t : t \in T\} \neq \emptyset$, because ${}^*M$ is saturated and thus contains a point $c'$. Then $\pi(c') \in \bigcap\{C_t : t \in T\}$ by the closedness of $C_t$, so (QC) follows. $\square$

**Remark 2.2.11.**

(i) The $\pi$-topology essentially depends on $\pi$; in particular, it depends on whether ${}^*M$ is saturated.

(ii) Note that we do not need the saturation of ${}^*M$ to establish the completeness of the $\pi$-topology.

### 2.2.1 Universal specialisations

**Definition 2.2.12.** For (partial) specialisation $\pi : {}^*M \to M$, we say that the pair $({}^*M, \pi)$ is **universal (over $M$)** if for any $M' \succeq {}^*M \succeq M$, any finite subset $A \subset M'$, and a specialisation $\pi' : A \cup {}^*M \to M$ extending $\pi$, there is an elementary embedding $\alpha : A \to {}^*M$, over $A \cap {}^*M$, such that

$$\pi' = \pi \circ \alpha \text{ on } A.$$

**Example 2.2.13.** In Examples 2.2.4 and 2.2.5, the pairs $({}^*\mathbb{C}, \pi)$ and $({}^*K, \pi)$ are universal if the structures (in the language extended by $\pi$) are $\omega$-saturated. Using A. Robinson's analysis of valuation theory (1977), M. Piatkus has shown that for Example 2.2.5 this is the case when the valuation in question is *maximal*, and in particular when $\Gamma = \mathbb{R}$.

**Example 2.2.14.** Let $\tilde{\mathbb{R}}$ be the compactified reals with Zariski topology. That is, the universe of $\tilde{\mathbb{R}}$ is $\mathbb{R} \cup \{\infty\}$, and $\mathcal{C}$ consists of Zariski closed subsets of Cartesian powers of this universe. Obviously, $\tilde{\mathbb{R}}$ is definably equivalent to the field of reals.

Consider an elementary extension $^*\tilde{\mathbb{R}} \succ \tilde{\mathbb{R}}$ and the surjective standard-part map $st :^* \tilde{\mathbb{R}} \to \tilde{\mathbb{R}}$. These are specialisations with regards to $\mathcal{C}$, because the metric topology extends the Zariski one.

Whatever the choice for $^*\tilde{\mathbb{R}}$, $st$ is not universal. Indeed, in some bigger elementary extension, choose $a \notin {}^*\tilde{\mathbb{R}}$. We may assume that $a > 0$. So, $a = b^2$ for some $b \in {}^*\tilde{\mathbb{R}}$. Because $a$ is transcendental over $^*\tilde{\mathbb{R}}$, the extension $\pi$ of $st$ sending $a$ to $-1$ is a partial specialisation (with regards to Zariski topology). However, we cannot embed $a$ (and so $b$) into $^*\tilde{\mathbb{R}}$ because $st(b)$ then would be a real number and $-1 = st(a) = st(b)^2$, a contradiction.

**Proposition 2.2.15.** *For any structure* M, *there exists a universal pair* ($^*$M, $\pi$). *If* M *is quasi-compact, then* $\pi$ *is total.*

*Proof.* We construct $^*$M and $\pi$ by the following routine process [compare with the proof of A.4.20(iii)]:

Start with any $M_0 \succeq M$ and a specialisation $\pi_0 : M_0 \to M$. We construct a chain of length $\omega$ of elementary extensions $M_0 \preceq M_1 \preceq \cdots \preceq M_i \cdots$ and partial specialisations $\pi_i : M_i \to M$, $\pi_i \supseteq \pi_{i-1}$. In case M is quasi-compact, $\pi_i$ is going to be total.

To construct $M_{i+1}$ and $\pi_{i+1}$, we first consider the set

$$\{\langle A_\alpha, \bar{a}_\alpha, p_\alpha(\bar{x})\rangle : \alpha < \kappa_i\}$$

of all triples where $A_\alpha$ is a finite subset of $M_i$, $\bar{a}_\alpha \in M^n$, and $p_\alpha(\bar{x})$ is an $n$-type over $M \cup A_\alpha$.

Now we construct specialisations $\pi_{i,\alpha}$, $\alpha \leq \kappa_i$, such that $\pi_{i,\alpha} \supset \pi_i$ and the domain of $\pi_{i,\alpha}$ is $N_{i,\alpha}$.

Let $N_{i,0} = M_i$ and $\pi_{i,0} = \pi_i$. At limit steps $\alpha$, we take the union. On successive steps, we follow this process:

If there is $\bar{b} \models p_\alpha$ and a specialisation $\pi' \supset \pi_{i,\alpha}$, sending $\bar{b}$ to $\bar{a}_\alpha$, let $N_{i,\alpha+1} = N_{i,\alpha} \cup \{\bar{b}\}$, $\pi_{i,\alpha+1} = \pi'$.

Otherwise, let $N_{i,\alpha} = N_{i,\alpha+1}$, $\pi_{i,\alpha+1} = \pi_{i,\alpha}$.

Now put $M_{i+1}$ as a model containing $N_{i,\kappa_i}$, and $\pi_{i+1} \supseteq \pi_{i,\kappa_i}$, a specialisation from $M_{i+1}$ to M. If M is quasi-compact, by Proposition 2.2.15 we assume $\pi_i$ is total.

It follows from the construction that for any $M' \succeq M_{i+1} \succeq M$, any finite $B \subset M'$, and a specialisation $\pi' : B \cup M_i \to M$ extending $\pi_i$, there is an elementary isomorphism $\alpha : B \to M_{i+1}$ over $M \cup (B \cap M_i)$ such that

$$\pi' = \pi \circ \alpha \text{ on } B.$$

Now take $^*M = \bigcup_{i<\omega} M_i$, $\pi = \bigcup_{i<\omega} \pi_i$. $\qquad \square$

**Remark 2.2.16.** One can easily adjust the construction to get *M saturated over M. The following lemma will be useful later:

**Lemma 2.2.17.** *There exists a structure* **M $\succeq$ *M *and a specialisation* $\pi^*$: **M $\to$ *M *such that*

(i) (**M, $\pi^*$) *is a universal pair over* *M.
(ii) (**M, $\pi \circ \pi^*$) *is a universal pair over* M.
(iii) **M *is* $|$*M$|^+$-*saturated.*

*Proof.* We build **M just like in the proof of Proposition 2.2.15, by taking care of (i) and (ii), respectively, in alternating steps. Namely, we build an ascending sequence of elementary models

$$*M \preceq M_1 \preceq M_2 \cdots \preceq M_n \preceq \cdots,$$

and an ascending sequence of specialisations $\pi_n^* : M_n \to$ *M, such that these conditions hold:

(i) For odd $n$, if $M' \succeq M_{n-1}$, $A \subseteq M_{n-1}$ is finite, $\bar{b} \in M'$, and $\pi' : M_{n-1} \cup \{\bar{b}\} \to$ *M is a specialisation extending $\pi_{n-1}^*$, then there is a map $\tau : \bar{b} \to M_n$, elementary over *M $\cup$ A, such that $\pi_n^* \tau(\bar{b}) = \pi'(\bar{b})$.
(ii) For even $n > 0$, if $M' \succeq M_{n-1}$, $A \subseteq M_{n-1}$ is finite, $\bar{b} \in M'$, and $\pi' : M_{n-1} \cup \{\bar{b}\} \to$ M is a specialisation extending $\pi \circ \pi_{n-1}^*$, then there is a map $\tau : \bar{b} \to M_n$, elementary over M $\cup$ A, such that $(\pi \circ \pi_n^*)\tau(\bar{b}) = \pi'(\bar{b})$.

We can now take **M $= \bigcup M_n$ and $\pi^* = \cup \pi_n^*$. By repeating the process, we can take **M to be $|$*M$|^+$-saturated. $\qquad\square$

## 2.2.2 Infinitesimal neighbourhoods

We assume from now on that, if not stated otherwise, $\pi$ is a universal specialisation.

**Definition 2.2.18.** For a point $a \in M^n$, we call an **infinitesimal neighbourhood of** $a$ the subset in *$M^n$ given as

$$V_a = \pi^{-1}(a).$$

Clearly then, for $a, b \in M$ we have $V_{(a,b)} = V_a \times V_b$.

**Definition 2.2.19.** Given $b \in M^n$, denote the $n$-type over *M as

$$\text{Nbd}_b(y) = \{\neg Q(c', y) : Q \in \mathcal{C}, M \models \neg Q(c, b), c' \in V_c, c \in M^k\}.$$

$$\text{Nbd}_b^0(y) = \{\neg Q(y) : Q \in \mathcal{C}, M \models \neg Q(b)\}.$$

As usual, $\mathrm{Nbd}_b(^*M)$ stands for the set of realisations of the type in $^*M$ and $\mathrm{Dom}\,\pi$, the domain of $\pi$ in $^*M$.

**Remark 2.2.20.** $\mathrm{Nbd}_b^0$ is just the restriction of the type $\mathrm{Nbd}_b$ to $M$.
Equivalently,

$$\mathrm{Nbd}_b^0(^*M) = \bigcap\{U(^*M) : U \text{ basic open}, b \in U\}.$$

**Lemma 2.2.21.**

(i)

$$V_b = \mathrm{Nbd}_b(^*M) \cap \mathrm{Dom}\,\pi.$$

(ii) *Given a finite $a'$ in $^*M$ and a type $F(a', y)$ over $a'$, $b' \in V_b$ exists satisfying $F(a', b')$, provided the type $\mathrm{Nbd}_b(y) \cup \{F(a', y)\}$ is consistent.*

*Proof.*

(i) If $b' \in V_b$ and $\neg Q(c', y) \in \mathrm{Nbd}_b(y)$, then necessarily $^*M \models \neg Q(c', b')$; otherwise, $M \models Q(c, b)$. Hence, $b' \in \mathrm{Nbd}_b(^*M)$.

   Conversely, suppose $b'$ realises $\mathrm{Nbd}_b$ in $^*M$ and $\pi(b') = c$. Then $c = b$, otherwise, $M \models \neg c = b$ requires $\neg c' = y \in \mathrm{Nbd}_b(y)$ for every $c' \in \pi^{-1}(c)$, including $c' = b'$, which contradicts the choice of $b'$.

(ii) Suppose the type is consistent; that is, there exists $b''$ realising $F(a', y)$ and $\mathrm{Nbd}_b(y)$ in some extension $M'$ of $^*M$. Then an extension $\pi' \supseteq \pi$ to $\mathrm{Dom}\,\pi \cup \{b''\}$, defined as $\pi'(b'') = b$, is a specialisation to $M$. Because $(^*M, \pi)$ is universal, we can find $b' \in {}^*M$ such that $\pi(b') = b$, and the type of $b'$ over $Ma'$ is equal to that of $b''$ over $Ma'$. (Here, the $A$ of the definition is equal to $a'b''$, $A \cap {}^*M = a'$, and $\alpha$ sends $b''$ to $b'$.) So, $F(a', b')$ holds. $\qquad\square$

In particular, the universality of the specialisation guarantees that the structure induced on $V_b$ is 'rich'; that is, $\omega$-saturated.

**Exercise 2.2.22.** *A definable subset $A \subseteq M^n$ is $C_\pi$-closed if and only if (iff) for every $a \in M^n$, $V_a \cap {}^*A \neq \emptyset$ implies $a \in A$; correspondingly, $A$ is $C_\pi$-open iff for every $a \in A$, $V_a \subseteq {}^*A$. Also, $\mathrm{Int}(A) = \{a \in A : V_a \subseteq {}^*A\}$.*

**Lemma 2.2.23.** *Let $S_0$ be a relatively closed subset of $S$. Suppose for some $s \in S$ there exists $s' \in V_s \cap S_0$. Then $s \in S_0$.*

*Proof.* We have by definition $S_0 = \bar{S}_0 \cap S$ and $s = \pi(s') \in \bar{S}_0$. $\qquad\square$

**Proposition 2.2.24.** *Assuming that $\pi$ is universal, a definable subset $S \subseteq M^n$ is closed in the sense of $C$ iff $S$ is $\pi$-closed.*

*Proof.* The left-to-right implication follows from Lemma 2.2.23. Now suppose $S$ is $\pi$-closed. Let $\bar{S}$ be the closure of $S$ in the $\mathcal{C}$-topology. Suppose $s \in \bar{S}$.

*Claim.* The type $\mathrm{Nbd}_s(y) \cup S(y)$ is consistent.

*Proof.* Suppose towards a contradiction, it is not. Then $^*\mathrm{M} \models S(y) \rightarrow Q(c', y)$ for some closed $Q$ such that $\pi(c') = c$ and $\mathrm{M} \models \neg Q(c, s)$.

Take any $t \in S$. Then $^*\mathrm{M} \models Q(c', t)$ and so $\mathrm{M} \models Q(c, t)$. This means that $S \subseteq Q(c, M)$. Hence $\bar{S} \subseteq Q(c, M)$, but this contradicts the fact that $s \notin Q(c, M)$. Claim proved.

It follows from the claim that $\mathcal{V}_s \cap {}^*S \neq \emptyset$. By Exercise 2.2.22, we get $s \in S$ and hence $S = \bar{S}$. $\qquad\square$

**Corollary 2.2.25.** *Suppose for a given topological structure* $\mathrm{M} = (M, \mathcal{C})$ *there exists a total universal specialisation* $\pi : {}^*M \rightarrow M$. *Then, the $\mathcal{C}$-topology satisfies* (P); *that is, it is closed under projections.*

The next theorem, along with facts proved previously, essentially summarises the idea that the approach via specialisations covers all the topological data in classical cases and is generally a more flexible tool when one deals with coarse topologies.

**Theorem 2.2.26.** *Let* $\mathrm{M}$ *be a topological structure. Let* $\mathrm{M} \prec {}^*\mathrm{M}$ *be an elementary extension saturated over* $M$.

*The following two conditions are equivalent:*

*(i) There is exactly one total specialisation* $\pi : {}^*\mathrm{M} \rightarrow \mathrm{M}$; *and*
*(ii)* $(M, \mathcal{C})$ *is a compact Hausdorff topological space.*

*Proof.* Assume (i). Then $\mathrm{M}$ is quasi-compact, and, using Proposition 2.2.15, $\pi$ is a universal specialisation. By Proposition 2.2.24, $\mathcal{C} = \mathcal{C}_\pi$. It remains to show that for any given $a \in M$, the intersection $\bigcap\{U : U \text{ open}, a \in U\}$ (also equal to $\mathrm{Nbd}_a^0(M)$) contains just the one point $a$. Suppose towards the contradiction that also $b$ is in this intersection and $b \neq a$. Take $a' \in \mathcal{V}_a \setminus \{a\}$. Then

$$a' \in \mathrm{Nbd}_a^0({}^*M) = \mathrm{Nbd}_b^0({}^*M).$$

Now define a new partial map $\pi' : a' \mapsto b$. This is a partial specialisation, because for every basic closed set $Q({}^*M)$ containing $a'$, $\neg Q$ does not define an open neighbourhood $U$ of $b$, so $b \in S(M)$. Extend $\pi'$ to a total specialisation by Proposition 2.2.7. This is different from $\pi$, contrary to our assumptions. We proved (ii).

Assume (ii). The existence of a specialisation $\pi$ is now given by Proposition 2.2.7. Then $\pi$ is unique because for any $a' \in {}^*M$, $\pi(a') = a$ iff $a' \in \mathrm{Nbd}_a^0({}^*M)$.  □

### 2.2.3 Continuous and differentiable function

**Definition 2.2.27.** Let $F(x, y)$ be the graph of a function on an open set $D$, $f : D \to M$. We say that $f$ is **strongly continuous** if $F$ is closed in $D \times M$, and for any universal specialisation $\pi : {}^*M \to M$ for every $a' \in \mathrm{Dom}\,\pi \cap D({}^*M)$ we have $f(a') \in \mathrm{Dom}\,\pi$.

**Remark 2.2.28.** If M is quasi-compact, then by Proposition 2.2.7, any function with a closed graph is strongly continuous.

**Exercise 2.2.29 (topological groups).** *Let G be a topological structure with a basic ternary relation P defining a group structure on G with the operation*

$$x \cdot y = z \equiv P(x, y, z)$$

*given by a closed P.*

*Suppose that G is quasi-compact. Consider* ${}^*G \succ G$, *a universal specialisation* $\pi : {}^*G \to G$ *and the infinitesimal neighbourhood* $\mathcal{V} \subseteq {}^*G$ *of the unit. Then* $\mathcal{V}$ *is a non-trivial normal subgroup of* $G^*$. *In particular,* ${}^*G$ *cannot be simple.*

**Proposition 2.2.30.** *Given strongly continuous functions* $f : D^n \to M$ *and* $g_i : U \to D$, $i = 1, \ldots, n$, *the composition*

$$f \circ g : x \mapsto f(g_1(x), \ldots, g_n(x))$$

*is strongly continuous.*

*Proof.* First we note that the graph of $f \circ g$ is closed. Indeed, by Proposition 2.2.24, it is enough to show that if $a' \in D({}^*M)$, $\pi(a') = a \in D(M)$ and $b' = f \circ g(a')$, $\pi(b') = b$, then $f \circ g(a) = b$. By definition $b' = f(c_1', \ldots, c_n')$, for $c_i' = g_i(a')$. By continuity, $c_i' \in \mathrm{Dom}\,\pi$, $\pi(c_i') = c_i$ and $g_i(a) = c_i$. By the same argument, $f(c_1, \ldots, c_n) = b$.  □

Observe also the following.

**Lemma 2.2.31.** *The graph of a strongly continuous function* $M^m \to M$ *on an irreducible M is irreducible.*

*Proof.* We claim that the graph $F$ of $f$ is homeomorphic to $M^m$ via the map

$$\langle x_1, \ldots, x_m \rangle \mapsto \langle x_1, \ldots, x_m, f(x_1, \ldots, x_m) \rangle;$$

that is, the image and inverse image of a closed subset are closed. The latter follows from axioms (L). The former follows by Proposition 2.2.24 and the assumption on $f$.

Thus, $F$ is irreducible. □

**Definition 2.2.32.** Assume that $M = K$ is a topological structure, which is an expansion of a field structure (in the language of Zariski-closed relations). We say that the function $f : K^m \to K$, $f = f(\bar{x}, y)$, $\bar{x} = \langle x_1, \ldots, x_{m-1} \rangle$, has **derivative with respect to** $y$ if there exists a strongly continuous function $g : K^{m+1} \to K$ and a function $f_y : K^m \to K$ with closed graph such that

$$g(\bar{x}, y_1, y_2) = \begin{cases} \frac{f(\bar{x}, y_1) - f(\bar{x}, y_2)}{y_1 - y_2}, & \text{if } y_1 \neq y_2 \\ f_y(\bar{x}, y_1), & \text{otherwise} \end{cases}$$

If this holds for $f$, we say that $f$ is **differentiable by** $y$.

**Example 2.2.33.** Consider the Frobenius map Frob : $y \mapsto y^p$ in a field $K$ of characteristic $p > 0$ and its inverse, Frob$^{-1}$. Both maps are strongly continuous because, for $\alpha \in \mathcal{V}_0$ (infinitesimal neighbourhood of 0), both $\alpha^p \in \mathcal{V}_0$ and $\alpha^{\frac{1}{p}} \in \mathcal{V}_0$. The derivative of Frob is the constant function 0 because for any $y \in {}^*K$, the element

$$\frac{\text{Frob}(y + \alpha) - \text{Frob}(y)}{\alpha} = \alpha^{p-1}$$

specialises to 0.

There is no derivative of Frob$^{-1}$ because

$$\frac{\text{Frob}^{-1}(y + \alpha) - \text{Frob}^{-1}(y)}{\alpha} = \alpha^{\frac{1-p}{p}}$$

specialises to no element in $K$.

**Proposition 2.2.34.** *Assume that $K$ is irreducible. The derivative $f_y$ is then uniquely determined.*

*Proof.* First, note that by assumptions $K^{m+1}$ is irreducible. By Lemma 2.2.31, the graph $G$ of $g$ is irreducible.

Suppose $\tilde{f}_y$ is also a derivative of $f$. The graph $\tilde{G}$ of the corresponding $\tilde{g}$ is irreducible then also, but the closed set $G \cap \tilde{G}$ contains the set

$$\left\{ \left\langle \bar{x}, y_1, y_2, \frac{f(\bar{x}, y_1) - f(\bar{x}, y_2)}{y_1 - y_2} \right\rangle : \bar{x} \in K^{m-1}, y_1, y_2 \in K, y_1 \neq y_2 \right\},$$

which is obviously open in $G$ and $\tilde{G}$. Thus, $G = \tilde{G}$. □

**Remark 2.2.35.** Assume that $+$, $\times$, $^{-1}$, and $f$ are strongly continuous in $K$, and $f$ is derivative with respect to $y$. Then by Proposition 2.2.30, for every infinitesimal $\alpha \in V_0$,

$$\pi : \frac{f(\bar{x}, y + \alpha) - f(\bar{x}, y)}{\alpha} \mapsto f_y(\bar{x}, y).$$

In other words, there exists $\beta \in V_0$ such that

$$\frac{f(\bar{x}, y + \alpha) - f(\bar{x}, y)}{\alpha} = f_y(\bar{x}, y) + \beta.$$

**Exercise 2.2.36.** *Assume that $+$, $\times$, $^{-1}$, $f$, and $h$ are strongly continuous and that $f$ and $h$ are differentiable in $K$. Then*

1. $(y^n)_y = ny^{n-1}$, *for any integer $n > 0$;*
2. *every polynomial $p$ is differentiable with respect to its variables;*
3. $(f + h)_y = f_y + h_y$;
4. $(f \cdot h)_y = f_y \cdot h + h_y \cdot f$; *and*
5. $f(\bar{x}, h(\bar{x}, y))_y = f_y(\bar{x}, h(\bar{x}, y)) \cdot h(\bar{x}, y)_y$.

In later sections, we use the notation $\frac{\partial f}{\partial y}$ interchangeably with $f_y$.

# 3

# Noetherian Zariski structures

Zariski geometries are abstract structures in which a suitable generalisation of Zariski topology makes sense. Algebraic varieties over an algebraically closed field and compact complex spaces in a natural language are examples of (Noetherian) Zariski geometries. The main theorem by Hrushovski and this author states that under certain non-degeneracy conditions, a one-dimensional Noetherian Zariski geometry can be identified as an algebraic curve over an algebraically closed field. The proof of the theorem exhibits a way to develop algebraic geometry from purely geometric abstract assumptions without involving any algebra at all.

## 3.1 Topological structures with good dimension notion

We introduce a dimension notion on sets definable in M. We are interested in the case when dimension satisfies certain conditions.

### 3.1.1 Good dimension

We assume that to any non-empty projective $S$, a non-negative integer called **the dimension of** $S$, $\dim S$, is attached.

We postulate the following properties of a good dimension notion:

(DP) **Dimension of a point** is 0,

(DU) **Dimension of unions:** $\dim(S_1 \cup S_2) = \max\{\dim S_1, \dim S_2\}$,

(SI) **Strong irreducibility:** For any irreducible $S \subseteq_{cl} U \subseteq_{op} M^n$ and its closed subset $S_1 \subseteq_{cl} S$, if $S_1 \neq S$, then $\dim S_1 < \dim S$,

(AF) **Addition formula:** For any irreducible $S \subseteq_{cl} U \subseteq_{op} M^n$ and a projection map pr $: M^n \to M^m$,

$$\dim S = \dim \mathrm{pr}\,(S) + \min_{a \in \mathrm{pr}\,(S)} \dim(\mathrm{pr}^{-1}(a) \cap S).$$

(FC) **Fibre condition:** For any irreducible $S \subseteq_{cl} U \subseteq_{op} M^n$ and a projection map pr $: M^n \to M^m$, there exists $V \subseteq_{op}$ pr $S$ (relatively open) such that

$$\min_{a \in \text{pr}(S)} \dim(\text{pr}^{-1}(a) \cap S) = \dim(\text{pr}^{-1}(v) \cap S), \qquad \text{for any } v \in V \cap \text{pr}(S).$$

More specifically, we say that dim is a good dimension for closed subsets if (OP) − (FC) hold for $U = M^n$.

**Remark 3.1.1.** Note that for Noetherian topological structures, we could have defined dimension of closed sets to be the Krull dimension, as in the work of Hrushovski and Zilber (1993); see also Section 3.3. This is quite convenient and in particular the strong irreducibility (SI) becomes automatic. However, this would disagree with some key natural examples, such as compact complex manifolds (see Section 3.4.2). Also, in the more general context of non-Noetherian structures, the reduction to irreducibles is quite subtle (see Section 6.1), and it is not even clear if the Krull dimension can work in this case.

Once a dimension is introduced, one can give a precise meaning to the notion of a generic point used broadly in a geometric context.

**Definition 3.1.2.** For $M \prec M'$,

$$S \subseteq_{cl} M^n \text{ is irreducible, and } a' \in S(M'). \tag{$*$}$$

We say that $a'$ is **generic in** $S$ if dim $S$ is the smallest possible among $S$ satisfying $(*)$.

### 3.1.2 Zariski structures

We use the following property to generalise the (P) of 2.1:

(SP) **Semi-properness** of projection mappings: Given a closed irreducible subset $S \subseteq_{cl} M^n$ and a projection map pr $: M^n \to M^k$, there is a proper closed subset $F \subset \overline{\text{pr } S}$ such that $\overline{\text{pr } S} \setminus F \subseteq \text{pr } S$.

**Definition 3.1.3.** Noetherian topological structures with good dimension notion for closed subsets satisfying (SP) are called **Zariski structures**, sometimes with the adjective *Noetherian,* to distinguish them from the analytic Zariski structures introduced later.

**Exercise 3.1.4.** *Prove that, for a closed $S \subseteq_{cl} M^n$, pr $S$ is constructible.*

In many cases we assume that a Zariski structure also satisfies these properties:

(EU) **Essential uncountability:** If a closed $S \subseteq M^n$ is a union of countably many closed subsets, then there are finitely many among the subsets, the union of which is $S$.

The following is an extra condition crucial for developing a rich theory for Zariski structures:

(PS) **Pre-smoothness:** For any closed irreducible $S_1, S_2 \subseteq M^n$, the dimension of any irreducible component of $S_1 \cap S_2$ is not less than

$$\dim S_1 + \dim S_2 - \dim M^n.$$

**Remark 3.1.5.** Note that (DCC) guarantees that $S$ is the union of irreducible components.

For simplicity, we add also the extra assumption that $M$ **itself is irreducible.** However, most of the arguments in the chapter hold without this assumption.

**Exercise 3.1.6.**

1. *In (FC) and (AF), we can write $S(a, M)$ instead of $\mathrm{pr}^{-1}(a) \cap S$, if $\mathrm{pr}$ is the projection on the first m coordinates.*
2. *Give an example where strict inequality may hold in Exercise 3.1.7.*
3. *If $S$ is a closed infinite set, then $\dim S > 0$.*
4. *$\dim M^k = k \cdot \dim M$.*
5. *$\dim S \leq \dim M^k$, for every constructible $S \subseteq M^k$.*
6. *Assume that M is compact. Let $S$ and $\mathrm{pr}\, S$ be closed, $\mathrm{pr}\, S$ be irreducible, and all the fibres $\mathrm{pr}^{-1}(a) \cap S$, $a \in \mathrm{pr}\, S$, be irreducible and of the same dimension. Then $S$ is irreducible.*

**Exercise 3.1.7.** *For a topological structure M with a good dimension and a subset $S \subseteq_{\mathrm{cl}} U \subseteq_{\mathrm{op}} M^n$, assume that there is an irreducible $S^0 \subseteq_{\mathrm{cl}} S$ with $\dim S^0 = \dim S$. Then*

$$\dim S \geq \dim \mathrm{pr}\,(S) + \min_{a \in \mathrm{pr}\,(S)} \dim(\mathrm{pr}^{-1}(a) \cap S)$$

*(i.e. a 'reducible' version of (AF)).*

*In particular, the reducible version of (AF) holds in Noetherian Zariski structures for any closed $S$.*

## 3.2 Model theory of Zariski structures

### 3.2.1 Elimination of quantifiers

**Theorem 3.2.1.** *A Zariski structure M admits elimination of quantifiers; that is any definable subset $Q \subseteq M^n$ is constructible.*

*Proof.* Recall that every Boolean combination of closed sets can be written in the form (3.2), Section 3.1.2.

We now let pr $: M^{n+1} \to M^n$ be the projection map along $(n+1)$-th coordinate. It is enough to prove that pr $(Q)$ is again of the form (3.2), if $Q \subseteq M^{n+1}$ is. Without loss of generality, we may assume that $Q = S \setminus P$ and is non-empty, and we may use induction on dim $S$.
Let

$$d_S = \min\{\dim S(a, M)) : a \in \text{pr}\, S\};$$

$$F = \{b \in \text{pr}\, S : \dim P(b, M) \geq d_S\}.$$

Let $\bar{F}$ be the closure of the set $F$. This is a proper subset of the closure $\overline{\text{pr}\, S}$, by (FC), and so dim $\bar{F} < \dim \overline{\text{pr}\, S}$, because $\overline{\text{pr}\, S}$ is irreducible (Exercise 2.1.4).

Let $S' = S \cap \text{pr}^{-1}(\bar{F})$. Because $\bar{F} \cap \text{pr}\, S \neq \text{pr}\, S$, we have $S' \subsetneq S$, and hence, because $S$ is irreducible, dim $S' < \dim S$.

Clearly,

$$\text{pr}\, Q = \text{pr}\,(S \setminus P) \subseteq \text{pr}\,(S' \setminus P) \cup (\text{pr}\, S \setminus F). \tag{3.1}$$

However, pr $S \setminus F \subseteq \text{pr}\, Q$, because if $b \in \text{pr}\, S \setminus F$, then $P(b, M) \subsetneq S(b, M)$ (i.e. $b \in \text{pr}\, Q$). So, equality holds in Equation (3.1). We can now apply induction to $S' \setminus P$ and use the fact that pr $S \setminus F$ is already in the desired form. $\qquad\square$

**Exercise 3.2.2.** *Let* M *be a Zariski structure,*

1. *Let* $C \subseteq M^k$ *be an irreducible set of dimension 1, and let* $S \subseteq C^m$ *be relatively closed and irreducible of dimension* $n > 1$. *Then*
   *(i)* $m \geq n$, *and*
   *(ii) There is a projection* pr $: C^m \to C^n$, *for some choice of n co-ordinates from m, and an open dense* $U \subseteq C^n$ *such that* $S^1 = \text{pr}^{-1}(U) \cap S$ *is dense in S and the projection map is finite-to-one on* $S^1$.
2. *Let* $C \subseteq M^k$ *be an irreducible set of dimension 1. Then C is strongly minimal [i.e. any definable (with parameters) subset of C is either finite or the complement to a finite subset].*

Let $\bar{Q}$ be **the closure** of a set $Q$, the smallest closed set containing $Q$.

**Remark 3.2.3.** Any constructible $Q$ has the form

$$Q = \cup_{i \leq k}(S_i \setminus P_i) \tag{3.2}$$

for some $k$, where $S_i$, $P_i$ are closed sets and $P_i \subset S_i$, $S_i$ is irreducible. Consequently

$$\bar{Q} = \bigcup_{i \leq k} S_i.$$

The dimension of a constructible set is

$$\dim Q := \dim \bar{Q} = \max_{i \le k} \dim S_i.$$

**Lemma 3.2.4.** *For a Zariski structure, the following form of the fibre condition holds:*

(FC') **Fibre condition:** *for any projection* pr *and a closed irreducible* $S \subseteq M^n$, *the set*

$$\mathcal{P}^{pr}(S, k) = \{a \in pr\, S : \dim(S \cap pr^{-1}(a)) > k\}$$

*is constructible and is contained in a proper (relatively) closed subset of* pr $S$, *provided* $k \ge \min_{a \in pr(S)} \dim(pr^{-1}(a) \cap S)$.

*Proof* (by induction on dim $S$). The statement is obvious for dim $S = 0$. It is also obvious for $k < k_0 = \min_{a \in pr(S)} \dim(pr^{-1}(a) \cap S)$, so we assume that $k \ge k_0$.

For dim $S = d > 0$, let $U$ be the open subset of pr $S$ on which $\dim(pr^{-1}(a) \cap S)$ is minimal. Then dim $U = \dim pr\, S$ because pr $S$ is constructible and irreducible. By (AF), the dimension of the set

$$S^0 = \bigcup_{a \in U} pr^{-1}(a) \cap S$$

is equal to dim $S$. It follows that the complement $S' = S \setminus S^0$ is of lesser dimension.

Note that (FC) in a Noetherian structure by Exercise (2.1.3) implies the fibre condition for arbitrary closed sets. It is clear that under our assumptions $\mathcal{P}^{pr}(S, k) = \mathcal{P}^{pr}(S', k)$. By induction, the latter is contained in a subset closed in pr $S \setminus U$, and therefore is closed in pr $S$. $\qquad\square$

**Remark 3.2.5.** We can also use a weakened form of this addition formula:

(AF') For any irreducible closed $S \subseteq M^n$ and a projection map pr $: M^n \to M^m$,

$$\dim S = \dim pr(S) + \min_{a \in pr(S)} \dim(pr^{-1}(a) \cap S).$$

One can check that this form of the addition formula is sufficient for proving elimination of quantifiers for Noetherian Zariski structures, by eventually restricting ourselves to (AF') we do not narrow the definition.

### 3.2.2 Morley rank

First, we give a model-theoretic interpretation of the *essential uncountability* property (EU).

**Lemma 3.2.6.** *A Zariski structure* M *satisfies (EU) iff it is* $\omega_1$-*compact, that is, all countable types are realised.*

*Proof.* The direction from right to left is immediate, shown by the compactness theorem. For the opposite direction, we have to check that any descending chain

$$Q_0 \supseteq Q_1 \supseteq \cdots Q_i \supseteq \cdots$$

of non-empty definable subsets of $M^n$ has a common point. We may assume that all $Q_i$ are of the same dimension and of the form $S \setminus P_i$ for closed $S$ and $P_i$. Now, apparently the intersection $\bigcap Q_i$ is non-empty iff $S \neq \bigcup P_i$, which follows immediately from (EU). □

**Remark 3.2.7.** Because $M$ may have an uncountable language, the notion of $\omega_1$-compact is weaker then the notion of $\omega_1$-saturated. R. Moosa (2005b) has shown that there are compact complex manifolds $M$ which are not $\omega_1$-saturated in the natural language. On the other hand, he proved that a compact complex $M$ is $\omega_1$-saturated if it is Kähler.

**Theorem 3.2.8.** *Any Zariski structure* M *satisfying (EU) is of finite Morley rank. More precisely,* rk $Q \leq$ dim $Q$ *for any definable set* $Q$.

*Proof.* We prove by induction on $n$ that rk $Q \geq n$ implies dim $Q \geq n$.

From definition, dim $Q = 0$ iff rk $Q = 0$ follows. Now, to prove the general case by a contradiction, suppose that, for some $Q$, $\dim(Q) \leq n > 0$ and $\mathrm{rk}\,(Q) \geq n + 1$. We may assume that $Q$ is irreducible. By the assumptions on Morley rank, for any $i \in \mathbb{N}$ there are disjoint $Q_1, \ldots Q_i$ with $\mathrm{rk}(Q_i) \geq n$ such that

$$Q \supseteq Q_1 \cup \cdots \cup Q_i.$$

Let $i = 2$. By the induction hypothesis, $\dim(Q_1) \geq n$ and $\dim(Q_2) \geq n$. By the irreducibility of $Q$, we then have $\dim(Q_1 \cap Q_2) \geq n$, which is a contradiction. □

## 3.3 One-dimensional case

We discuss here a specific axiomatisation of one-dimensional Zariski structures introduced by Hrushovski and Zilber (1993). It is more compact and easier to use in applications (such as in Hrushovski, 1996). Using Theorem 3.2.8, it is apparent that every one-dimensional pre-smooth Zariski structure (with irreducible universe) is strongly minimal and satisfies the definition given by Hrushovski and Zilber (1993), given below. It is not clear if the inverse holds, but we show that all the properties postulated in Section 3.1.2, except perhaps (FC), follow from (Z1)–(Z3). The main classification theorem, 4.4.1, is proved without use of (FC), and effectively only (Z1)–(Z3) is needed [see also the proof given by Hrushovski and Zilber (1993)].

**Definition 3.3.1.** A **one-dimensional Zariski geometry** on a set $M$ is a Noetherian topological structure satisfying the properties (Z1)–(Z3) and with dimension defined as a Krull dimension; that is, the dimension of a closed irreducible set $S$ is the length $n$ of a maximal chain of proper closed irreducible subsets

$$S_0 \subset \cdots \subset S_n = S.$$

Dimension of an arbitrary closed set is the maximum dimension of its irreducible components.

Note that by this definition (SI) holds; that is, irreducibles are strongly irreducible.

The axioms for one-dimensional Zariski geometry are

(Z1) $\mathrm{pr}\, S \supseteq \overline{\mathrm{pr}(S)} \setminus F$, for some proper closed $F \subseteq_{\mathrm{cl}} \overline{\mathrm{pr}(S)}$.

(Z2) For $S \subseteq_{\mathrm{cl}} M^{n+1}$, there is $m$ such that for all $a \in M^n$, $S(a) = M$ or $|S(a)| \leq m$.

(Z3) $\dim M^n \leq n$. Given a closed irreducible $S \subseteq M^n$, every component of the diagonal $S \cap \{x_i = x_j\}$ $(i < j \leq n)$ is of dimension $\geq \dim S - 1$.

**Lemma 3.3.2.** $M^k$ *is irreducible.*

*Proof.* By induction on $k$. Note that $M$ is irreducible because every infinite closed subset $T \subseteq M$ must be equal to $M$ by (Z2), with $n = 1$ and $S = \{a\} \times T$.

Let $S_1 \cup S_2 = M^{k+1}$ and $S_1, S_2 \subseteq_{\mathrm{cl}} M^{k+1}$. Consider

$$S_i^* = \{a \in M^n : a^\frown x \in S_i, \text{ for all } x \in M\}.$$

Clearly $S_i^* \subseteq M^n$ is closed, and for any $a \in M^k$, $S_i(a) = \{x \in M : a^\frown x \in S_i\}$ is closed and $M = S_1(a) \cup S_2(a)$. So, $S_i(a) = M$ for some $i$. Thus, $S_1^* \cup S_2^* = M^k$, so by induction $S_i^* = M^k$, for some $i$, and $S_i = M^{k+1}$. □

**Lemma 3.3.3.** *Let $S_1 \subseteq M^k$ and $S_2 \subseteq M^m$ be both irreducible. Then*

*(i) $S_1 \times S_2$ is irreducible;*

*(ii) $\dim(S_1 \times S_2) \geq \dim S_1 + \dim S_2$.*

*Proof.*

(i) Let pr be the projection $M^{k+m} \to M^m$. By definition, $\mathrm{pr}(S_1 \times S_2) = S_2$. Suppose $S_1 \times S_2 = P_1 \cup P_2$, and $P_1, P_2 \subseteq_{\mathrm{cl}} M^k + m$. For each $a \in S_1$, $S_2 = P_1(a) \cup P_2(a)$, so $P_1(a) = S_2$ or $P_2(a) = S_2$. Consider $P_i^* = \{a \in M^k : a^\frown x \in P_i \text{ for all } x \in S_2\}$. These are closed subsets of $S_1$, and $S_1 = P_1^* \cup P_2^*$. By irreducibility, $S_1 = P_i^*$ for some $i$; hence, $S_1 \times S_2 = P_i$.

(ii) Let $S_1^0 \subset \cdots \subset S_1^{d_1} = S_1$ and $S_2^0 \subset \cdots \subset S_2^{d_2} = S_2$ be maximal length chains of irreducible closed subsets of $S_1$ and $S_2$, so that $\dim S_1 = d_1$ and $\dim S_2 = d_2$ by definition. Now

$$S_1^0 \times S_2^0 \subset \cdots \subset S_1^0 \times S_2^{d_2} \subset S_1^1 \times S_2^{d_2} \subset S_1^2 \times S_2^{d_2} \subset \cdots \subset S_1^{d_1} \times S_2^{d_2}$$

is the chain of closed irreducible subsets of $S_1 \times S_2$ of length $d_1 + d_2$. □

**Lemma 3.3.4.** *Let* $P \subseteq_{\mathrm{cl}} M^n$ *and* $S \subseteq_{\mathrm{cl}} P \times M$. *Suppose* $S(a)$ *is finite for some* $a \in P$. *Then* $S(a)$ *is finite for all* $a \in P \setminus R$ *for some proper closed subset* $R \subseteq_{\mathrm{cl}} P$.

*Proof.* If $S(a)$ is infinite, then $S(a) = M$. It suffices to prove that $\{a : S(a) = M\}$ is closed, but this is the intersection of the sets $\{a : a^\frown b \in S\}$, for all $b \in M$, which is closed. □

The next statement is just (PS), which generalises (Z3).

**Lemma 3.3.5.** *Suppose* $S_1, S_2 \subseteq M^n$ *are both irreducible,* $\dim S_i = d_i$. *Then every irreducible component of* $S_1 \cap S_2$ *is at least of dimension* $d_1 + d_2 - n$.

*Proof.* Let $D_i$ be the diagonal $x_i = x_{n+i}$ in $M^{2n}$, and let $D = \bigcup_i D_i$. There is an obvious homeomorphism between $S_1 \cap S_2$ and $(S_1 \times S_2) \cap D$. By Lemma 3.3.3, $S_1 \times S_2$ is irreducible and at least of dimension $d_1 + d_2$. Hence, it suffices to show that $\dim S \cap D \geq \dim S - n$ for any irreducible closed subset $S$ of $M^{2n}$. This follows by applying (Z3) to the intersection with diagonals $D_i$, $i = 1, \ldots, n$, in succession. □

**Lemma 3.3.6.** *Let* $S \subseteq_{\mathrm{cl}} M^n$ *be irreducible,* $\mathrm{pr} : M^n \to M^k$.

  (i) *If* $\overline{\mathrm{pr}\, S} = M^k$, *then* $\dim S \geq k$.
  (ii) *If* $\mathrm{pr}^{-1}(a) \cap S$ *is finite and non-empty for some* $a$, *then* $\dim S \leq k$.
  (iii) *If* $\dim S = k$ *iff there exists* $\mathrm{pr}$ *as in (a) and (b).*

*Proof.*

  (i) By induction on $k$, we have $\mathrm{pr}\, S \supseteq M^k \setminus F$ for some proper closed $F \subseteq M^k$. For $a \in M$, let $F(a) = \{y \in M^{k-1} : a^\frown y \in F\}$. If for all $a \in M$, $F(a) = M^{k-1}$, then $F = M^k$, which is a contradiction. Choose $a \in M$ so that

$F(a) \subset M^{k-1}$, which is proper. Let $S' = \{x \in S : \mathrm{pr}\, x \in \{a\} \times M^{k-1}\}$. This is a closed subset of $S$.

Let $\mathrm{pr}^- : M^k \to M^{k-1}$ be the projection, forgetting the first co-ordinate, $\mathrm{pr}^+ = \mathrm{pr}^-\mathrm{pr}$. Clearly $\overline{\mathrm{pr}^+ S'}$ contains $M^{k-1} \setminus F(a)$. Because $M^{k-1}$ is irreducible, the closure $\overline{M^{k-1} \setminus F(a)}$ contains $M^{k-1}$, so by Exercise 2.1.4 (ii), $\overline{\mathrm{pr}^+ S''} = M^{k-1}$ for some component $S''$ of $S'$. By induction, $\dim S'' \geq k - 1$, so $\dim S \geq k$.

(ii) Pick $a \in M^k$ such that $\mathrm{pr}^{-1}(a) \cap S$ is finite. Clearly $\mathrm{pr}^{-1}(a) \cap S = (\{a\} \times M^{n-k}) \cap S$. So by Lemma 3.3.5,

$$0 = \dim \mathrm{pr}^{-1}(a) \cap S \geq \dim S + \dim(\{a\} \times M^{n-k}) - n = \dim S - k.$$

Hence, $\dim S \leq k$.

(iii) We first prove

*Claim.* Let $S \subset M^n$ be irreducible, proper, and closed. There is a projection $\mathrm{pr}_1 : M^n \to M^{n-1}$ and $a \in M^{n-1}$ such that $\mathrm{pr}_1^{-1}(a) \cap S$ is finite.

*Proof* (by induction on $n$). Let $\mathrm{pr} : M^n \to M^{n-1}$ be the projection along the last co-ordinate. If the projection has a finite fibre on $S$, then we are done, so we assume otherwise, that is, $\mathrm{pr}^{-1}(a) \cap S$ is an infinite closed subset of $\{a\} \times M$ for all $a \in \mathrm{pr}\, S$. By (Z2), the fibre is equal to $\{a\} \times M$. Hence,

$$\mathrm{pr}\, S = \{a \in M^{n-1} : \forall x \in M\ a^\frown x \in S\},$$

which is a closed subset of $M^{n-1}$, and $S = \mathrm{pr}\, S \times M$. So, $\mathrm{pr}\, S$ is a proper subset of $M^{n-1}$. By induction, there is a projection $\mathrm{pr}' : M^{n-1} \to M^{n-2}$ with a finite fibre on $\mathrm{pr}\, S$. The projection $x^\frown y \mapsto \mathrm{pr}'x^\frown y$ satisfies the requirement.

Now it follows from (b) that $\dim M^n \leq n$. Hence, $\dim M^n = n$.

Finally, we prove (c) by induction on $n$. In case $S = M^n$, we take $\mathrm{pr}$ to be the identity map. If $S$ is proper, then by the claim there exists a projection $\mathrm{pr}_1 : M^n \to M^{n-1}$ and $a \in M^{n-1}$ such that $\mathrm{pr}_1^{-1}(a) \cap S$ is finite. Let $P = \overline{\mathrm{pr}_1(S)}$, which is irreducible by Exercise 2.1.4. By induction, there exists $\mathrm{pr}_2 : M^{n-1} \to M^k$ such that $\mathrm{pr}_2 P$ is dense in $M^k$ and $\mathrm{pr}_2^{-1}(a) \cap P$ is finite for some $a \in \mathrm{pr}_2 P$. We have $\dim P = k$. Using (Z1) and Lemma 3.3.4, there is $F \subset P$ closed such that $\mathrm{pr}_1(S) \supseteq P \setminus F$ and $\mathrm{pr}_1^{-1}(b) \cap S$ is finite for all $b \in P \setminus F$. Because $\dim F < k$, by (a) $\mathrm{pr}_2(F)$ is not dense in $M^k$. Choose $a \in M^k \setminus \overline{\mathrm{pr}_2(F)}$ such that $\mathrm{pr}_2^{-1}(a) \cap P$ is finite. Then $(\mathrm{pr}_2\mathrm{pr}_1)^{-1}(a) \cap S$ is finite, but $\mathrm{pr}_2\mathrm{pr}_1(S)$ contains $\mathrm{pr}_2(P) \setminus \mathrm{pr}_2(F)$ and hence is dense in $M^k$. Thus, $\mathrm{pr}_2\mathrm{pr}_1$ is the projection satisfying (a) and (b). $\qquad\square$

We now draw some model-theoretic conclusions from this theory. Moreover, we use essentially general model theory (the theory of strongly minimal structures) outlined in Appendix B to prove that the dimension notion is good in this sense.

The following statements are special cases of Theorems 3.2.1 and 3.2.8, but we do not use (FC) and (AF) in the proofs.

**Proposition 3.3.7.** *Let* M *be a one-dimensional Zariski geometry. The theory of* M *admits elimination of quantifiers. In other words, the projection of a constructible set is constructible.*

*Proof.* We must show that if $S \subseteq M^n \times M$ is a closed subset, $F \subseteq S$ is a closed subset, and if $\mathrm{pr}_1$ denotes the projection to $M^n$, then $\mathrm{pr}_1(S \setminus F)$ is a Boolean combination of closed sets. We show this by induction on $\dim S$. Note that we can immediately reduce to the case where $S$ is irreducible. Let $S_1 = \overline{\mathrm{pr}_1 S}$. Then $S_1$ is irreducible, and for some proper closed $H \subseteq S_1$, $\mathrm{pr}_1 S \supseteq S_1 \setminus H$. Let $S_0 = \{x \in M^n : \forall y \, x ^\frown y \in S\}$ and $F_0 = \{x \in M^n : \forall y \, x ^\frown y \in F\}$. Then $S_0$ and $F_0$ are closed and $S_0 \subseteq S_1$. The case $S_0 = S_1$ is trivial, because then $\mathrm{pr}_1(S \setminus F) = S_0 \setminus F_0$. Let $F_1 = \overline{\mathrm{pr}_1(F)}$. If $F_1$ is a proper subset of $S_1$, then so is $F_2 = F_1 \cup H$, and $(F_2 \times M) \cap S$ is a proper subset of $S$ and hence has smaller dimension. Thus by induction, $\mathrm{pr}_1((F_2 \times M) \cap C \setminus F)$ is a Boolean combination of closed sets. So is $\mathrm{pr}_1(S \setminus F) = \mathrm{pr}_1((F_2 \times M) \cap S \setminus F) \cup (S_1 \setminus F_2)$. The remaining case is $S_1 = F_1$, $S_0 \neq F_0$. In this case, we claim that $S = F$. Because $S$ is irreducible, it suffices to show that $\dim S = \dim F$. In fact, $\dim S = \dim S_1$ and $\dim F = \dim F_1$. This follows from the characterisation of $\dim F$ in Lemma 3.3.6. $\square$

**Corollary 3.3.8.**

(i) M *is strongly minimal (see Definition B.1.22).*

(ii) *For* $S \subseteq M^n$ *constructible, the Morley rank of* $S$ *is equal to the dimension of the closure of* $S$, $\mathrm{rk}\, S = \dim \overline{S}$.

*Proof.*

(i) Let $E \subseteq M^n \times M$ be a denable set. It is enough to show that $E(a)$ is finite or co-finite, with a uniform bound for all $a \in M^n$. We may take $E = S \setminus F$, with $S$ and $F$ closed. If $S(a)$ is finite, then $E(a)$ is finite with the same bound. The result is immediate from (Z2) applied to $S$ and to $F$.

(ii) We use the properties of Morley rank for strongly minimal structures Lemma B.1.26. It is enough to prove the statement that $S$ is irreducible. Let $P = \overline{S}$. We use induction on $\dim P$. Clearly, for $\dim P = 0$, $S$ is finite and so $\mathrm{rk}\, S = 0$.

Let $\dim P = k$. By Lemma 3.3.6, there is an open subset $S' \subseteq P$ and an open subset $T \subseteq M^k$ such that a projection $\mathrm{pr} : M^n \to M^k$ is a finite-to-one surjective map $S' \to T$. We have $\dim M^k \setminus T < k$, and so $\mathrm{rk}\, M^k \setminus T < k$.

So, rk $T = k$, by Lemma B.1.26(i)–(iii). Now, Lemma B.1.26(vi) implies rk $S' =$ rk $T$. Because also by induction rk $P \setminus S' = \dim P \setminus S' < k$, we have rk $P = k$. $\qquad\square$

**Remark 3.3.9.** In particular, the addition formulas for dimension in the Lemma B.1.26(v)–(vi) hold.

For a constructible set $S$, dim $S$ will stand for dim $\overline{S}$.

**Proposition 3.3.10.** *Suppose (FC) holds for* M. *Then the addition formula (AF) is true.*

*Proof.* Consider an irreducible $S \subseteq_{\mathrm{cl}} U \subseteq_{\mathrm{op}} M^n$ and a projection map pr : $M^n \to M^m$. Denote $d = \min_{a \in \mathrm{pr}(S)} \dim(\mathrm{pr}^{-1}(a) \cap S)$. Let $V$ be the open set as stated in (FC). Then the open subset $S' = \mathrm{pr}^{-1}(V) \cap S$ is of dimension dim pr $S + d$, by the addition formula Lemma B.1.26(v) for Morley rank. However, dim $S =$ dim $S'$, which proves (AF). $\qquad\square$

# 3.4 Basic examples

## 3.4.1 Algebraic varieties and orbifolds over algebraically closed fields

Let $K$ be an algebraically closed field and $M$ the set of $K$-points of an algebraic variety over $K$. We are going to consider a structure on $M$:

**The natural language for algebraic varieties,** $M$, is the language with the basic $n$-ary relations $\mathcal{C}$, given by the Zariski-closed subsets of $M^n$.

**Theorem 3.4.1.** *Any algebraic variety M over an algebraically closed field in the natural language and the dimension notion (as that of algebraic geometry) is a Zariski structure. The Zariski structure is complete if the variety is complete. It satisfies (PS) if the algebraic variety is smooth. It satisfies (EU) iff the field is uncountable.*

*Proof.* Use a book on algebraic geometry, e.g. those by Shafarevich (1997) or Danilov (1994). (L) and (DCC) follow immediately from the definition of a Zariski structure and the Noetherianity of polynomial rings (Sections 1 and 2, Chapter 1 of Shafarevich's book. The irreducible decomposition is discussed in Section 3. (SP) is Theorem 2 and (PS) is Theorem 5 of Section 5 of the same book.) The fibre condition (FC), along with the addition formula (AF), is given by *fibres dimension theorem* in Danilov's work (1994). $\qquad\square$

**Orbifolds.** Consider an algebraic variety $M$ and the structure $M$ in the natural language as before on the set of its $K$-points. Suppose there is a finite group $\Gamma$ of regular automorphisms acting on $M$. We consider the set of orbits $M/\Gamma$ and the canonical projection $p : M \to M/\Gamma$. Define the *natural topological structure with dimension (orbifold)* on $M/\Gamma$ to be given by the family $\mathcal{C}_\Gamma$ of subsets of $(M/\Gamma)^n$, all $n$, which are of the form $p(S)$ for $S \subseteq M^n$ closed in $M$. Set $\dim p(S) := \dim S$. In Section 3.7, we prove Proposition 3.7.22: *The orbifold $M/\Gamma$ is a Zariski structure. The orbifold is pre-smooth if $M$ is.*

Note that generally $M/\Gamma$ is not an algebraic variety even if $M$ is one.

Using the general quantifier-elimination Theorem, 3.2.1, we have a corollary.

**Corollary 3.4.2.**

(i) *Any definable subset of $K^n$, an algebraically closed field $K$, is a Boolean combination of affine varieties (zero-sets of polynomials).*

(ii) *Any definable subset of $M^n$, for an orbifold M over an algebraically closed field, is a Boolean combination of Zariski-closed subsets.*

### 3.4.2 Compact complex manifolds

For the definitions and references on complex manifolds, we refer mainly to Gunning and Rossi (1965). As in that book, we identify theorems from Gunning and Rossi by a triplet consisting of a Roman number, a letter, and an Arabic number. Chapter V, Section B, Statement 20, for example, will be V.B.20, and in case the reference to a book is omitted, we mean Gunning and Rossi's 1965 book.

**The natural language for a compact complex manifold** $M$ has the analytic subsets of $M^n$ as basic $n$-ary relations $\mathcal{C}$.

**Theorem 3.4.3.** *Any compact complex manifolds $M$ in the natural language and dimension given as complex analytic dimension is a complete Zariski structure and satisfies assumptions (PS) and (EU).*

*Proof.* We need to check the axioms:

(L) is given in definitions.

(P) is Remmert's theorem, V.C.5.

(DCC): To see this, first notice that any analytic S is at most a countable union of irreducible analytic $S_i$ and the cover $S = \bigcup_i S_i$ is locally finite (Grauert and Remmert, 1984; A.3, decomposition lemma). By compactness, the number of irreducible components is finite. Now, (DCC) for compact analytic sets follows from (DCC) for irreducible ones, which is a consequence of axiom (SI). The latter [as well as (DU)] is immediate in III.C.

The condition (AF) is the second part of Remmert's theorem (V.C.5), which also states that the minimum dimension of fibres is achieved on an open subset.

(FC) is less immediate.*

Let U be a neighbourhood of a point $b \in S$, which is locally biholomorphic to a complex disk of dimension $r_n = \dim M^n$, and $S \cap U$ is given as the zero-set of $f_1, \ldots, f_m$ holomorphic in $U$. Projection pr is given by holomorphic functions $g_1, \ldots, g_{r_{n-1}}$. Then $p^{-1}(a) \cap S$ is the zero-set of $f_1, \ldots, f_m$ and $g_1 = a_1, \ldots, g_{r_{n-1}} = a_{r_{n-1}}$, where $a_i$ are the coordinates of $a \in \mathrm{pr}(S \cap U)$. If $\langle a, b \rangle$ is a point in $a$-fibre, then the fact that the dimension of the fibre $> k \geq \min \dim$ of fibres implies

$$\text{the rank of Jacobian of } (f, g) \text{ in } b \text{ is less than } r - k. \qquad (*)$$

This condition is equivalent to vanishing of all (r-k)-minors of the Jacobian, so it is a (local) analytic condition on $\langle a, b \rangle$. Let $S'$ be the global analytic set defined by $(*)$ in every $U$. By the construction, all components of dimension greater than k of fibres $\mathrm{pr}^{-1}(a) \cap S$ lie in $S'$. This gives $\mathcal{P}(S', k) = \mathcal{P}(S, k)$. By the choice of $k$, there is $\langle a, b \rangle$ not in $\mathcal{P}(S, k)$, and thus $S'$ is a proper subset of $S$. Now the induction by $\dim S$ finishes the proof of (FC).

(EU) is given by V.B.1.

(PS): Let $S_1$, $S_2$ be irreducible subsets of $M^n$. It is easy to see that $S_1 \cap S_2$ is locally biholomorphically isomorphic to $S_1 \times S_2 \cap \mathrm{Diag}(M^n \times M^n)$. Now notice that locally $M^n$ is represented by disks of $\mathbb{C}^d$, where $d = \dim M^n$. Now the condition (PS) is satisfied in $M$ by III.C.11, because the diagonal is given by $d$ equations; each of them decreases the dimension at most by 1. $\qquad \square$

**Remark 3.4.4.** In fact, the theorem holds for compact analytic spaces [in Gunning and Rossi's sense (1965)], except for the pre-smoothness condition.

**Remark 3.4.5.** Proposition 3.7.22 about orbifolds is true also in the context of complex geometry, that is, when M is a complex manifold.

Again from Theorem 3.2.1 one derives the following corollary.

**Corollary 3.4.6.** *The family of constructible subsets of a compact complex manifold is closed under projections.*

---

*Alternately, you can use Theorem 9F, p. 240, in Whitney's book on complex analytic varieties. You need to add to it Remmert's proper mapping theorem.

### 3.4.3 Proper varieties of rigid analytic geometry

Now we consider a less classical subject, the *rigid analytic geometry* (see Bosch *et al.*, 1984, for references). It is built over a completion of a non-Archimedean-valued algebraically closed field $K$. The main objects are analytic varieties over $K$. The natural language for an analytic variety $M$ is again the language with analytic subsets of $M^n$ as basic relations. We have to warn the reader that here the definition of a neighbourhood and so of an analytic subset is much more involved than in the complex case. The main obstacle for an immediate analogy is the fact that the non-Archimedean topology on $K$ is highly disconnected.

**Theorem 3.4.7.** *Let $M$ be a proper (rigid) analytic variety. Then $M$, with respect to the natural language, is a complete Zariski structure satisfying (EU). It is pre-smooth if the variety is smooth.*

*Proof.* For references, we use Bosch *et al.* (1984) with the appropriate enumeration of the statements.

(L) follows from definitions.

Notice that $M^n$ is proper by Lemma 9.6.2.1.

The projections

$$\mathrm{pr} : M^{n+1} \to M^n$$

are proper by Proposition 9.6.2.4 and particularly the subsequent comments.

If $S \subseteq M^{n+1}$ is analytic, then $\mathrm{pr}(S) \subseteq M^n$ is analytic by the proper mapping theorem, 9.6.3.3. This gives us (P).

Because analytic subsets are defined locally and locally are in a correspondence with Noetherian coordinate rings, which are also factorial domains, we have (DCC) locally. By the properness, we can reduce any admissible open covering to a finite one; thus, we have (DCC) globally, that is, for analytic subsets of $M^n$. Also, the notion of irreducibility has a local ring-theoretic representation.

The dimension is defined locally as the Krull dimension. We thus get (SI), (DU), (DP), (FC), and (AF).

To prove (EU), assume $S \subseteq M^n$ is an analytic subvariety and $S = \bigcup_{i \in \mathbb{N}} S_i$, a union of analytic varieties. Because local coordinate rings of $S$ are factorial, locally $S$ decomposes into finitely many irreducible analytic subvarieties. Consider an open admissible subset $U \subseteq M^n$, where we may assume $S \cap U$ is irreducible. Because $S \cap U$ is a complete metric space, at least one of $S_i \cap U$ must be a first category subset; that is, it must contain an open subset of $S \cap U$. It follows that $\dim S_i = \dim S$, and thus by irreducibility $S_i \cap U = S \cap U$. This proves that (EU) holds locally. By the properness again, any admissible open covering can be assumed to be finite, and this yields that $S$ is a union of finitely many $S_i$'s.

Finally, by the same reason as in the complex case, if $M$ is smooth, we have (PS). □

**Exercise 3.4.8.** *Let* $\Lambda \subseteq \mathbb{C}^n$ *be the additive subgroup with an additive basis* $\{a_1, \ldots, a_{2n}\}$, *linearly independent over the reals. Then the quotient space* $T = \mathbb{C}^n / \Lambda$ *has a canonical structure of a complex manifold, called a* **torus**.

1. *Prove that in the natural language a commutative group structure is definable in* $T$. *Moreover, the group operation* $x \cdot y$ *and the inverse* $x^{-1}$ *are given by holomorphic mappings.*
2. *Prove, using literature, that if* $a_1, \ldots, a_{2n}$ *are algebraically independent and* $n > 1$, *then any analytic subset of* $T^n$ *is definable in the group structure* $(T, \cdot)$ *(the generic torus).*
3. *Prove that for a generic torus* $T$ *the classical analytic dimension is* $\dim T = n > 1$ *and at the same time* $T$ *is strongly minimal; that is, its Morley rank and the Krull dimension both equal 1.*

The reader interested in model-theoretic aspects of the theory of compact complex spaces may also consult the works by Moosa (2005a, 2005b), Aschenbrenner *et al.* (2006), and Pillay and Scanlon (2002, 2003).

### 3.4.4 Zariski structures living in differentially closed fields

The first-order theory $DCF_0$ of differentially closed fields of characteristic zero is one of the central objects of present-day model theory. This theory has quantifier elimination and is $\omega$-stable, which makes it model-theoretically nice (Marker, 2002), and at the same time its structure is **very reach.**

E. Hrushovski found an amazing application of the theory of differentially closed fields in combination with the classification theorem of Zariski geometries (see the main theorem, Section 4.4). He proved the Mordell–Lang conjecture for function fields (1996) and then extended his method to give a proof of another celebrated number-theoretic conjecture, that of Manin and Mumford.

In particular, Hrushovski used the fact (proved by him and Sokolovic) that any strongly minimal substructure $M$ of a model of $DCF_0$ is a Zariski structure. By removing a finite number of points, one can also make $M$ pre-smooth.

A. Pillay extended this result and proved it in 2002.

**Theorem 3.4.9 (A. Pillay).** *Let* $K$ *be a differentially closed field of characteristic zero, and let* $X \subseteq K^n$ *be a definable set of finite Morley rank and Morley* *degree* 1. *Then, after possibly removing from X a set of Morley rank smaller than X, X can be equipped with a pre-smooth Zariski structure D*, dim, *in such a way that the subsets of $X^n$ definable in K are precisely those definable in $(X; D)$.*

Of course, if we choose $K$ to be $\omega_1$-saturated, then $D$ also satisfies (EU).

## 3.5 Further geometric notions

### 3.5.1 Pre-smoothness

We assume here that M is a Zariski structure. We give a wider notion of pre-smoothness, which is applicable to constructible subsets in Zariski structures.

**Definition 3.5.1.** A constructible set $A$ will be called **pre-smooth** (with $M$) if for any relatively closed irreducible $S_1, S_2 \subseteq A^k \times M^m$, any irreducible component of the intersection $S_1 \cap S_2$ is of dimension not less than

$$\dim S_1 + \dim S_2 - \dim(A^k \times M^m).$$

We also discuss the following strengthening of (PS):

(sPS) M will be called **strongly pre-smooth** if for any constructible irreducible $A \subseteq M^r$ there is a definable pre-smooth $A_0 \subseteq A$ open in $A$.

**Definition 3.5.2.** A Zariski structure satisfying (sPS) and (EU) is called a **Zariski geometry**.

In Subsection 3.6.4, we show that a one-dimensional uncountable Zariski structure satisfying (PS) is a Zariski geometry.

**Lemma 3.5.3.** *Let A be an irreducible pre-smooth (with M) set, and let $S \subseteq A^k \times M^l$ be closed irreducible. Let pr be a projection of S on any of its co-ordinates, and assume that $\mathrm{pr}(S) = A^k$ and $r = \min_{a \in \mathrm{pr}(S)} \dim S(a, M)$. Then for every $a \in A^k$, every component of $S(a, M)$ has dimension not less than r. In particular, if $\dim S(a, M) = r$, then all components of $S(a, M)$ have dimension r.*

*Proof.* Take $a \in A^k$. Then by the pre-smoothness of $A$ (and Exercise 2.1.5(2)), every component $C$ of $S(a, M)$ satisfies

$$\dim C \geq \dim S + \dim(\mathrm{pr}^{-1}(a) \cap (A^k \times M^l)) - \dim(A^k \times M^l)$$
$$\geq \dim S - \dim \mathrm{pr}(S) = r. \qquad (3.3)$$

$\square$

**Definition 3.5.4.** For $D$, a definable set, and $b \in D$, define $\dim_b D$ the **local dimension of D** in $b$ to be the maximal dimension of an irreducible component of $D$ containing $b$.

**Corollary 3.5.5.** *A definable set A is pre-smooth (with M) iff for any $k$, $m$, and relatively closed sets $S_1, S_2 \subseteq A^k \times M^m$*

$$\dim_x(S_1 \cap S_2) \geq \dim_x(S_1) + \dim_x(S_2) - \dim(A^k \times M^m)$$

*for any $x \in S_1 \cap S_2$.*

*Proof.* Clearly, if the condition is satisfied, then $A$ is pre-smooth. For the converse, consider the components of $S_1$ and $S_2$ which contain $x$ and have maximal dimension in $S_1$ and $S_2$, respectively. $\qquad\qquad$ □

**Corollary 3.5.6.** *If A is irreducibly pre-smooth, then any open subset B of it is pre-smooth too.*

*Proof.* In the previous notations, let $x$ be an element of $(B^k \times M^m) \cap S_1 \cap S_2$, and let $U \subseteq A^k \times M^m$ be an irreducible component of $S_1 \cap S_2$ containing $x$ and of dimension equal to $\dim_x(S_1 \cap S_2)$. Then $U' = U \cap (B^k \times M^m)$ is an open, and hence dense, subset of $U$. So, $\dim U' = \dim U$. Also, $B^k \times M^m$ is a dense open subset of $A^k \times M^m$. It follows

$$\dim U' \geq \dim_x(S_1 \cap (B^k \times M^m)) + \dim_x(S_2 \cap (B^k \times M^m)) - \dim(B^k \times M^m).$$

$\qquad\qquad$ □

**Exercise 3.5.7.** *Let D be pre-smooth (with M).*

1. *If $D_1$ is open and dense in $D$, then $D_1$ is pre-smooth. [Show also that the following variations fail:*
   *If $D_1$ open and dense in $D$ and $D_1$ is pre-smooth, then $D$ is.*
   *If $\bar{D}_1 = D$ ($D$ pre-smooth), then $D_1$ is pre-smooth.]*
2. *Let F be a relatively closed, irreducible subset of $D \times M^l$, $\mathrm{pr}(F) = D$, and let*

   $$\hat{F} = \{(x, y_1, \ldots, y_k) : M \models F(x, y_1) \& \cdots \& F(x, y_k)\}.$$

   *Then every irreducible component of $\hat{F}$ has dimension $d$, where $d = \dim D + k(\dim F - \dim D)$.*
3. *If $F \subseteq D^m$ is irreducible of dimension $l$, then every component of $F^k$ is of dimension $k \cdot l$.*

The examples that follow demonstrate why we need in (2) to assume that $D$ is pre-smooth.

**Example 3.5.8.** We are going to clarify here the notion of pre-smoothness for algebraic curves over an algebraically closed field $K$. One has first to precisely explain the way an algebraic curve is considered a Zariski structure. Of course, the best of all is to consider it in the natural language as in Theorem 3.4.1 (i.e. all the analytic structure). Here, for simplicity, we use a different representation. It follows from the main theorem and later results that the present representation is equivalent to the natural one. So, speaking of an algebraic curve $C$, we consider the algebraic variety $M = C \times K$ over the algebraically closed field $K$ in the language containing all the algebraic subvarieties of $M^n$ (in the complete version we can consider an embedding of such an $M$ into $\mathbf{P}^n$, or $M = C \times \mathbf{P}^n$, for some $n$).

**Proposition 3.5.9.** *Let $C$ be an (irreducible) algebraic curve over an algebraically closed field $K$ and $\{a_1, \ldots a_m\}$ be the set of all singular points of $C$. Then*

(i) *there is a smooth algebraic curve $A$ and a regular finite-to-one mapping $f : A \to C$, Such that $f$ is a biregular bijection on $C \setminus \{a_1, \ldots a_m\}$;*
(ii) *$C$ is pre-smooth iff $f$ is a bijection.*

*Proof.*

(i) This is the classic 'removal of singularities'; for curves, it is given by Theorem 6, Section 5, Chapter II of Shafarevich's work (1977), combined with Theorem 3 in the same section.
(ii) We assume $A \subseteq \mathbf{P}^n \subseteq M$. If $f$ is a bijection, then it is a Z-homeomorphism, i.e. it maps Z-closed subsets of $C^n$ to that of $A^n$ and conversely. Hence, it transfers pre-smoothness from $A$ to $C$.

If $f$ is not a bijection, then $f^{-1}(a_i) = \{b_1^i, \ldots, b_{k_i}^i\}$, $k_i > 1$ for some of the $a_i$'s in $C$. Take

$$F_1, F_2 \subseteq C \times M \times M,$$

defined as

$$\langle x, y, z \rangle \in F_1 \text{ iff } x = f(y), \quad \langle x, y, z \rangle \in F_2 \text{ iff } x = f(z).$$

Now, $\dim F_1 = \dim F_2 = 1 + \dim M$, and the irreducible components of $F_1 \cap F_2$ are $\{\langle x, y, z \rangle : y = z \ \& \ x = f(y)\}$ and $\{\langle a_i, b_j^i, b_l^i \rangle\}$, for distinct $i, j \leq k$. Thus, some of the components are of dimension 0, whereas $\dim F_1 + \dim F_2 - \dim(C \times M) = 1$, contradicting pre-smoothness. $\square$

**Example 3.5.10.**
**The pre-smooth case:** Let $C_1 \subseteq K \times K$ be the projective curve

$$\{(x, y) : y^2 = x^3\}.$$

The point $(0, 0)$ is singular, yet $C_1$ is pre-smooth by the previous proposition, because the regular map $f : K \to C_1$, given by $t \mapsto (t^2, t^3)$, is a bijection.
**The non-pre-smooth case:** Let $C_2 \subseteq K \times K$ be the curve defined by

$$y^2 = x^3 + x^2.$$

The point $(0, 0)$ is singular on the curve. Consider the map $f : K \to C_2$ given by $t \mapsto (t^2 - 1, t(t^2 - 1))$. The $f$ is a bijection on $K \setminus \{1, -1\}$, but $f(1) = f(-1) = (0, 0)$; hence, by the previous proposition, $C_2$ is not pre-smooth.

### 3.5.2 Coverings in structures with dimension

In this section, M is a topological structure with dimension, and $(^*M, \pi)$ is a universal pair as constructed previously such that $^*M$ is saturated over M.

**Definition 3.5.11.** Assume $F(x, y) \subseteq_{\mathrm{cl}} V \subseteq_{\mathrm{op}} M^n \times M^k$ is irreducible and pr $: M^n \times M^k \to M^n$, $\mathrm{pr}(F) = D$ (thus $D$ should be irreducible too). We say then $F$ **is an (irreducible) covering of** $D$.

**Definition 3.5.12.** We call the number $r = \min_{a \in D} \dim F(a, M)$ *the dimension of a generic fibre.*
Then $a \in D$ is called **regular for** $F$ if $\dim F(a, y) = r$. The set of points regular for $F$ are denoted $\mathrm{reg}(F/D)$.

**Lemma 3.5.13.** $\dim(D \setminus \mathrm{reg}(F/D)) \leq \dim D - 2$.

*Proof.* The set of irregular points $F' = \{\langle a, b \rangle \in F : a \in (D \setminus \mathrm{reg}(F/D))\}$ is a proper closed subset of $F$. By (SI), $\dim F' < \dim F$. By Exercise 3.1.7,

$$\dim F' \geq r + 1 + \dim(D \setminus \mathrm{reg}(F/D)).$$

The required inequality follows. $\qquad\square$

**Corollary 3.5.14.** *Suppose $F$ is a covering of an irreducible $D$, $\dim D = 1$. Then, every $a \in D$ is regular for the covering.*

**Definition 3.5.15.** Let $F$ be an irreducible covering of $D$, $a \in D$. We say that $F$ is a **discrete covering of** $D$ at $a$ [or in $(a, b)$] if $\dim F(a, v) = 0$.
We say that $F$ is a **finite covering** at $a$ if $F(a, M)$ is finite.

Clearly, if $F$ is a finite covering of $D$ at $a$, then the dimension of a generic fibre of $F$ is 0. Namely, $a \in \text{reg}(F/D) = \{d \in D : \dim F(d, v) = 0\}$ and hence every $a' \in V_a$ is in $\text{reg}(F/D)$.

### 3.5.3 Elementary extensions of Zariski structures

We aim to show here that an elementary extension of a Zariski structure can be canonically endowed with a topology and a dimension notion so that it becomes a Zariski structure again.

**Definition 3.5.16.** For any $M' \succeq M$, introduce the notion of a closed relation in $M'$ by declaring the sets (relations) of the form $S(a, M'^m)$ closed for $S$, a closed $(l + m)$-ary relation in M, and $a \in M'^l$. The closed sets which are defined using parameters from a set $A$ are called $A$-closed.

To define dimension in $M'$, notice first that by (AF), if $S \subseteq M^{l+m}$ is M-closed, then there is a bound $m \dim M$ on the dimension of the fibres of $S$; hence, for every $a \in \text{pr}(S)$ there is a maximal $k \in \mathbf{N}$ such that $a \in \mathcal{P}(S, k)$. Because $\mathcal{P}(S, k)$ is a definable set, we define

$$\dim S(a, M') = \max\{k \in \mathbf{N} : a \in \mathcal{P}(S, k)\} + 1.$$

**Exercise 3.5.17.** *Let* $M' \succeq M$ *and S be a closed relation. Show that*

1. *for* $a \in M$, $\dim S(a, M) = \dim S(a, M')$;
2. *if* $S_1$ *is another closed relation and* $a', a'_1 \in M'$ *are such that* $S(a', M') = S_1(a'_1, M')$, *then* $\dim S(a', M') = \dim S_1(a'_1, M')$; *and*
3. *for any closed* $S_1, S_2$ *closed in M and any* $a'_1, a'_2$ *in* $M'$,

    $$\dim(S_1(a'_1, M') \cup S_2(a'_2, M')) = \max\{\dim S_1(a'_1, M'), \dim S_2(a'_2, M')\},$$

    *that is,* (DU) *holds in* $M'$.

So, the dimension notion is well defined.

**Definition 3.5.18.** For $a \in M'^n$, $A \subseteq M'$, define the **locus of** $a$ **over** $A$, locus $(a/A)$, to be the intersection of all $A$-closed sets containing $a$. Define the **(combinatorial) dimension of a tuple** $a$ **over** $A$

$$\text{cdim}\,(a/A) := \dim(\text{locus}(a/A)).$$

Clearly, for the canonical example of Zariski structure, an algebraically closed field $K$, for $a \in K^m$, $A \subseteq K'$,

$$\text{cdim}\,(a/A) = \text{tr.d.}(a/A).$$

The following lemma will be very useful throughout.

**Lemma 3.5.19.** *Assume that $S$ is a closed set in $M^{k+l}$, pr $: M^{k+l} \to M^k$ is the projection map, and $\overline{\text{pr } S} = \text{locus}(a/M)$ for some $a \in M'^k$. Then*

$$\dim S(a, M') = \min\{\dim S(a', M) : a' \in \text{pr}(S)\}.$$

*Proof.* Let $l = \dim S(a, M')$. Then $a \in \mathcal{P}(S, l - 1)$, and hence $\mathcal{P}(S, l - 1) = \text{pr } S$. $\qquad\Box$

By the definition of dimension in M', the lemma holds even when we take $a' \in M'$.

**Exercise 3.5.20.** *If for a specialisation $\pi : a' \mapsto a$, then*

$$\dim S(a', M') \le \dim S(a, M).$$

**Lemma 3.5.21.** *Let $S(x, y), S^1(x, y) \subseteq M^{k+l}$ be 0-closed, $a' \in M'$ and $S^1(a', M') \subset S(a', M')$, $\dim S^1(a', M') = \dim S(a', M')$. Then there is a closed $S^2(x, y)$ such that $S(a', M') = S^1(a', M') \cup S^2(a', M')$ and $S(a', M') \ne S^2(a', M')$.*

*Proof.* Without loss of generality, $S^1 \subseteq S$. Let $L = \text{locus}(a'/M)$ and $T = S(x, y) \cap \text{pr}^{-1}(L)$, $T^1 = S^1(x, y) \cap \text{pr}^{-1}(L)$, where pr $: M^{k+l} \to M^k$ is the projection map. Notice that $\overline{\text{pr}(T)} = \overline{\text{pr}(T^1)} = L$ and $S(a', M') = T(a', M'), S^1(a', M') = T^1(a', M')$. In the following argument, we can only use the fact that M itself is a Zariski structure.

Let $S^2$ be the union of all components $K$ of $T$ such that $K \not\subseteq T^1$. Clearly, $T(a', M') = T^1(a', M') \cup S^2(a', M')$. It is left to see that $S^2(a', M') \ne T(a', M')$.

Assume that $S^2(a', M') = T(a', M')$. Let $d = \dim T(a', M')$. Then there is a component $K$ of $T$, $K \not\subseteq T^1$, such that $\dim(K(a', M') \cap T^1(a', M')) = d$ [use (DU) and Exercise 3.5.17(3)]. But $a' \in \text{pr}(K \cap T^1)$, and hence $\overline{\text{pr}(K)} = \text{pr}(K \cap T^1) = L$. By Lemma 3.5.19 (applied in M') and Exercise 3.1.7 (applied in M), we have

$$\dim(K \cap T^1) \ge d + \dim L = \dim K,$$

which implies that $K \subseteq T^1$, contradicting our assumption. $\qquad\Box$

**Corollary 3.5.22.** M' *satisfies* (SI).

**Lemma 3.5.23.** M' *satisfies* (AF).

*Proof.* Let $S$ be a closed subset of $M^{r+n+k}$, $a' \in M'^r$, and $S(a', M', M')$ be an irreducible closed subset of $M'^{n+k}$. Let pr $: M'^{r+n+k} \to M'^{r+n}$ be the projection map, $L = \overline{\text{pr } S}$, and $S(a', b', M')$ be a fibre of the projection of a minimal possible dimension when $b' \in L(a', M')$, $d = \dim S(a', b', M')$. We want to prove that $\dim S(a', M', M') = \dim L(a', M') + d$.

W.l.o.g., we may assume that $S$ is irreducible. Also, by denoting $\mathrm{pr}_1 :$ $M^{r+n} \to M^r$, the projection on the $a$-co-ordinates, we may assume

$$\overline{\mathrm{pr}_1 L} = \mathrm{locus}(a'/M).$$

Then by Lemma 3.5.19 $\dim L(a', M')$ and $\dim S(a', M', M')$ are of minimal dimension among the fibres with parameters ranging in $\mathrm{pr}_1 L$ in $M'$.

Let

$$d_0 = \min\{\dim S(\langle a, b \rangle, M) : \langle a, b \rangle \in L\}.$$

Claim: $d_0 = d$.

Indeed, $d_0 \leq d$ by definition. To see the converse, suppose towards a contradiction that $d_0 < d$. Then, by (FC) there exists a proper $L' \subsetneq L$ closed in $L$ such that, for any given $a''$ in $M$ and $b' \in L(a'', M)$, if $\dim S(a'', b', M) > d_0$, then $\langle a'', b' \rangle \in L$. In particular, if $a''$ is such that $\dim S(a'', b', M) > d_0$ for all $b' \in L(a'', M)$, then $L(a'', M) \subseteq L'(a'', M)$. By elementary equivalence, this holds in $M'$, so $L(a', M') \subseteq L'(a', M')$. But $\dim \mathrm{pr}_1 L'$ has to be strictly less than $\dim \mathrm{pr}_1 L$ because $\dim L' < \dim L$. This contradicts the fact that $\mathrm{locus}(a'/M) \subseteq \overline{\mathrm{pr}_1 L'}$ and proves the claim.

We get by (AF), for sets in $M$,

$$\dim S = \dim \mathrm{pr}_1 L + \dim S(a', M', M')$$

[using the facts that $\dim S(a', M', M')$ is equal to the minimum of $\dim S(a, M, M)$ and that $\mathrm{pr}_1 L = \mathrm{pr}_1 \mathrm{pr}\, S$] and

$$\dim S = \dim L + d = \dim \mathrm{pr}_1 L + \dim L(a', M') + d.$$

(Here $\dim L(a', M')$ stands in place of the minimum of $\dim L(a, M)$ for $a \in \mathrm{pr}_1 L$.)

Hence,

$$\dim S(a', M', M') = \dim L(a', M') + d.$$

$\square$

**Lemma 3.5.24.** *Any descending sequence of closed sets in $M'$,*

$$S_1(a_1, M') \supset S_2(a_2, M') \supset \cdots \supset S_n(a_n, M') \supset \cdots$$

*stabilises at some finite step.*

*Proof.* Suppose not. We can express in a form of a countable type about $x_1^0$ the (infinite) statement that for each $n$ there are $x_2, \ldots, x_n$ such that

$$S_1(x_1^0, M) \supsetneq S_2(x_2, M) \supseteq \cdots \supsetneq S_n(x_n, M).$$

This type is consistent and is realised in M by an element $a_1^0$.

It follows from this definition of $a_1^0$ that the type about $x_2^0$ stating the existence, for each $n$, of $x_3, \cdots, x_n$ with

$$S_1(a_1^0, M) \supsetneq S_2(x_2^0, M) \supsetneq S_3(x_3, M) \supseteq \cdots \supsetneq S_n(x_n, M)$$

is also consistent. So, we can get $a_2^0$ in M for $x_2^0$. Continuing in this way, we get a strictly descending chain of closed sets in M, thus contradicting (DCC) in M. $\qquad\square$

**Theorem 3.5.25.** *For any essentially uncountable Noetherian Zariski structure* M, *its elementary extension* M' *with closed sets and dimension as defined previously is a Noetherian Zariski structure. If we choose* M' *to be* $\omega_1$-*compact, then it satisfies* (EU).

*Proof.* (L) and (SP) are immediate. (DU) is proved in Exercise 3.5.17(3). For (SI), use Lemma 3.5.21. (DP) and (FC) are immediate from the definition of dimension. (AF) and (DCC) have been proved in Lemmas 3.5.23 and 3.5.24. (EU) is a direct consequence of $\omega_1$-compactness. $\qquad\square$

Notice that the proposition fails in regard to (DCC) without assuming (EU) for M.

**Example 3.5.26.** Consider a structure M in a language with an only binary predicate $E$ defining an equivalence relation with finite equivalence classes – one class of size $n$ for each number $n > 0$. The first-order theory of the structure obviously has quantifier elimination. We declare closed all subsets of $M^k$ defined by positive Boolean combinations of $E$ and $=$, with parameters in $M$. This notion of closed satisfies (L), (P), and (DCC). Let dim $M = 1$, dim $E = 1$, dim $E(a, M) = 0$, for all $a \in M$. One can easily extend this to the dimension notion of any closed subset so that (DU) – (AF) and (PS) are satisfied, but (EU) obviously fails because $M$ is countable.

Notice that, given $a \in M$ and $b_1, \ldots, b_m \in E(a, M)$, we have also $E(a, M) \setminus \{b_1, \ldots, b_m\}$ as a closed set because it is a union of finitely many singletons.

Now consider an elementary extension M' of M with at least one infinite equivalence class, $E(a, M')$. By our definitions, the sets of the form $E(a, M') \setminus \{b_1, \ldots, b_m\}$ are closed in M', obviously contradicting (DCC).

Notice that this example shows also that (EU) is essential in Theorem 3.2.8. Indeed, the correct Morley rank of the set $M$ [calculated in an $\omega$-saturated model of Th($M$)] is 2, but dim $M = 1$.

**Example 3.5.27.** Consider the structure $(\mathbb{N}, <)$, the natural numbers with the ordering. The elementary theory of this structure is very simple, and one can easily see that any formula in free variables $v_1, \ldots, v_n$ is equivalent to a Boolean combination of formulas of the form

$$v_i \leq v_j \text{ and 'the distance between } v_i \text{ and } v_j \text{ is less than } n\text{'}.$$

Take a finite conjunction of the basic formulas to be closed in $(\mathbb{N}, <)$, set dim $N^k = k$, and extend the notion of dimension in an obvious way to all closed sets. One can check that this is a Zariski structure satisfying also the pre-smoothness condition, but not (EU). No elementary extension of the structure is Zariski.

Moreover, in contrast to Theorem 3.2.8, the Morley rank of $N$ in the structure is $\infty$ and the theory of the structure is unstable.

**Proposition 3.5.28.** *Suppose $D$ is an irreducible set in a Zariski structure* M *and $\pi : {}^*M \to M$ is a universal specialisation. Let* dim $D = d$ *and $b \in D$. Then there is $b' \in \mathcal{V}_b \cap D({}^*M)$, such that* cdim $(b'/M) = d$.

*Proof.* By Lemma 2.2.21, it is sufficient to show that the type $\{D(y)\} \cup \text{Nbd}_b(y)$ is consistent and can be completed to a type of dimension $d$ (i.e. every definable set in the completion is of dimension at least $d$).

Assume that this fails. Then there is a closed set $Q$ as in the type $\text{Nbd}_b$ such that dim$(\neg Q(c', y) \,\&\, D(y)) < d$. Then dim $Q(c', y) \,\&\, \bar{D}(y) \geq d$ and by irreducibility applied in ${}^*M$ (see Lemma 3.5.21), $D({}^*M) \subseteq Q(c', {}^*M)$. Hence $\models Q(c', b)$, and so by applying $\pi$ one gets $\models Q(c, b)$, contradicting the choice of $Q$.                                              $\square$

Recall that for $D$ (a definable relation) and $b \in D$, dim$_b D$ is the maximal dimension of an irreducible component $D$ containing $b$.

**Corollary 3.5.29.**

$$\text{dim}_b D = \max\{\text{cdim}\,(b'/M) : b' \in \mathcal{V}_b \cap D^*\};$$

*that is, dimension is a **local property**.*

**Lemma 3.5.30.** *The statement of Lemma 3.5.3 remains true in any* M$' \succ$ M. *That is, let $A$ be an irreducible pre-smooth set definable in* M *and $S \subseteq A^k \times M^l$ is closed irreducible. Let* pr *be a projection of $S$ on any of its co-ordinates, and assume that* pr $S = A^k$ *and $r = \min_{a \in \text{pr}(S)} \dim S(a, M)$. Then for every*

$a' \in A^k(M')$, every component of $S(a', M')$ has dimension not less than $r$. In particular, if $\dim S(a', M') = r$, then all components of $S(a', M')$ have dimension $r$.

*Proof.* The estimate (3.3) remains valid. Indeed, suppose towards the contradiction it is not. Then

$$S(a', M') = P_0(b'_0, M') \cup P_1(b'_1, M')$$

with $P_i$ 0-definable closed sets, $b'_i$ tuples in $M'$, and $\dim P_0(b'_0, M') < r$ and $P_0(b'_0, M') \nsubseteq P_1(b'_1, M')$. Then by elementary equivalence, there are $a$, $b_0$, and $b_1$ in $M$ such that $a \in A^k$, $S(a, M) = P_0(b_0, M) \cup P_1(b_1, M)$, $\dim P_0(b_0, M) < r$, and $P_0(b_0, M) \nsubseteq P_1(b_1, M)$, which clearly contradicts Lemma 3.5.3. □

Notice the similarity between the last part of the previous fact and Exercise 3.5.33 (1). (In Exercise 3.5.33, we do not assume pre-smoothness, though.)

**Proposition 3.5.31.** *If* $M' \succeq M$ *and* M *satisfies (sPS), then so does* M'.

*Proof.* Let $C$ be definable irreducible in $M^{r+n}$, $c \in M'^r$, and $C(c, M'n)$ is an irreducible subset of $M'^n$. By assumptions, there is an open subset $C_0$ of $C$ which is pre-smooth, and then $C_0(c, M'^n)$ is an open subset of $C(c, M'^n)$.

Let $S_1$ and $S_2$ be definable subsets of $M^{r+k}$ and $M^{s+k}$, respectively. Assume that for $a \in M'^r$ and $b \in M'^s$, the sets $S_1(a, M'^k)$ and $S_2(b, M'^k)$ are irreducible closed subsets of $C_0(c, M'^n)$. Let

$$T_1(c_1, M'^k) \cup \cdots \cup T_m(c_m, M'^k)$$

be the irreducible decomposition of $S_1(a, M'^k) \cap S_2(b, M'^k)$. By introducing mock variables, we can assume $a = b = c = c_1 \cdots = c_m$.

Without loss of generality, $S_1$, $S_2$, and all $T_i$ are irreducible and $\mathrm{locus}(a/M) = \mathrm{pr}(S_1) = \mathrm{pr}(S_2) = \mathrm{pr}(T_i) = \mathrm{pr}(C) = L$.

Using the fact that $M \preceq M'$, choose an $a' \in L$ in $M$ such that

$$S_1(a', M) \cap S_2(a', M) = T_1(a', M) \cup \cdots \cup T_m(a', M).$$

By Lemma 3.5.19, all the fibres $S_i(a', M)$, $T_j(a', M)$, and $C(a', M)$ $(i = 1, 2; \; j = \{1, \ldots, m\})$ are of the same dimension as $S_i(a, M)$ $T_j(a, M)$, and $C(a, M)$.

By Lemma 3.5.30, irreducible components of each of the sets $S_i(a', M)$, $T_j(a', M)$, and $C(a', M)$ are of the same dimension as the sets themselves.

By (sPS) for any $x \in S_1(a', M) \cap S_2(a', M)$,

$$\dim_x T_i(a', M) \geq \dim_x S_1(a', M) + \dim_x S_2(a', M) - \dim(C(a', M) \times M^k),$$

Thus,

$$\dim T_i(a, M') \geq \dim S_1(a, M') + \dim S_2(a, M') - \dim(C(a, M') \times M^k).$$

$\square$

**Exercise 3.5.32.** *Let* M *be a Zariski structure.*

1. *If $S$ is an $A$-closed irreducible set in $M'$, $a \in M'$, and then $S = \mathrm{locus}(a/A)$ iff $a$ is generic in $S$ over $A$.*
2. *If $Q$ is $A$-definable and $a$ is generic in $\bar{Q}$ over $A$, then $a' \in Q$.*
3. *$L = \mathrm{locus}(a^\frown b/A)$ iff $\overline{\mathrm{pr}\,(L)} = \mathrm{locus}(a/A)$ and $L(a, M) = \mathrm{locus}(b/Aa)$. Also, for every generic $a \in \overline{\mathrm{pr}\,(L)}$, there is $b$ such that $(a, b)$ is generic in $L$.*
4. *Let $S$ be an $A$-definable irreducible set, and assume that $a, b$ are generic in $S$ over $A$. Then $tp(a/A) = tp(b/A)$.*
5. *If $a$ is in the (model-theoretic) algebraic closure of $b \in M^l$ over $A$, then $\mathrm{cdim}\,(a/A) \leq \mathrm{cdim}\,(b/A)$.*

**Exercise 3.5.33.** *Let* M *be a Zariski structure, $a, b$ be tuples from* M, *and $A \subseteq M$.*

1. *If $L = \mathrm{locus}(a^\frown b/A)$, then the irreducible components of $L$ are of equal dimension. Also, the irreducible components of $L(a, M)$ are of equal dimension.*
2. *Prove the dimension formula:*

$$\mathrm{cdim}\,(a^\frown b/A) = \mathrm{cdim}\,(a/Ab) + \mathrm{cdim}\,(b/A).$$

## 3.6 Non-standard analysis

### 3.6.1 Coverings in pre-smooth structures

The following proposition states that some useful definable sets are very well approximated by closed sets when one assumes pre-smoothness. We also assume (DCC), though the latter can be omitted as we show later in the treatment of analytic Zariski structures.

**Proposition 3.6.1.** *Suppose $F \subseteq D \times M^k$ is an irreducible covering of a pre-smooth $D$ and $Q(z, y) \subseteq M^{n+k}$ closed. Define*

$$L(z, x) = \{(z, x) \in M^n \times D : Q(z, M^k) \supseteq F(x, M^k)\},$$

*and assume that the projection of $L$ onto $D$, $\exists z L(z, x)$, is dense in $D$.*

*Then there is a (relatively) closed* $\hat{L}(z, x) \subseteq M^n \times D$ *and* $D' \subset D$, dim $D' <$ dim $D$, *such that*

$$\hat{L} \cap \left(M^n \times \mathrm{reg}\,(F/D)\right) \subseteq L \subseteq \hat{L} \cup (M^n \times D').$$

*Proof.* Let $L_1, \ldots, L_d$ be all the irreducible components of $L$ for which $\exists z L_i(z, x)$ is dense in $D$. [Hence, dim $\exists z(L(z, x) \setminus \bigcup L_i(z, x)) < $ dim $D$.]

Let $\bar{L}_i(z, x)$ be the closure in $M^n \times D$ of $L_i$. Consider

$$\begin{aligned} S_i &= \{\langle z, x, y \rangle \in M^n \times D \times M^k : \bar{L}_i(z, x) \& F(x, y)\} \\ &= (\bar{L}_i \times M^k) \cap (M^n \times F). \end{aligned}$$

Let $S_i^0(z, x, y)$ be an irreducible component of $S_i$ such that

$$\exists z \exists y (S_i^0) \cap \mathrm{reg}\,(F/D) \neq \emptyset.$$

If we let $r$ be the dimension of a generic fibre of $F$, then for every $\langle c', a' \rangle$ generic in the projection $\exists y S_i^0$, we have dim $S_i^0(c', a', M) \leq r$. By (AF),

$$\dim \exists y S_i^0 + r \geq \dim S_i^0,$$

but by pre-smoothness,

dim $S_i^0$

$$\begin{aligned} &\geq (\dim \bar{L}_i + \dim M^k) + (\dim M^n + \dim F) - (\dim M^n + \dim D + \dim M^k) \\ &= \dim \bar{L}_i + (\dim F - \dim D) \\ &= \dim \bar{L}_i + r. \end{aligned}$$

Thus, dim $\exists y S_i^0 \geq \dim \bar{L}_i$ and by definitions $\exists y S_i^0 \subseteq \bar{L}_i$ and $\bar{L}_i$ irreducible, $\exists y S_i^0$ is dense in $\bar{L}_i$.

Let $\langle c', a', b' \rangle$ be generic in $S_i^0$. Because $\langle c', a' \rangle$ is generic in $\bar{L}_i$, it must be in $L_i$ because, since $F(a', b')$ holds, we have $Q(c', b')$. Because $S_i^0$ is irreducible and $Q$ is closed, we get

$$S_i^0 \subseteq \{(z, x, y) : Q(z, y) \,\&\, D(x)\}.$$

By the choice of $S_i^0$, we then have

$$S_i \subseteq \{(z, x, y) : Q(z, y) \vee x \in F \setminus \mathrm{reg}\,(F/D)\}.$$

Assume now that $\bar{L}_i(c, a)$ holds and that $a \in \mathrm{reg}(F/D)$. Then for every $b$ with $F(a, b)$, we have $S_i(c, a, b)$, and hence, by the above, $Q(c, b)$. Thus, we proved $\bar{L}_i(c, a)$ implies $L_i(c, a)$ or dim $F(a, M) > r$.

Take now $\hat{L} = \bar{L}_1 \vee \cdots \vee \bar{L}_d$ and

$$D' = \left\{ x : \dim F(x, v) > r \vee \exists z \left( L(z, x) \setminus \bigcup_{i \leq d} L_i(z, x) \right) \right\}.$$

These $\hat{L}$ and $D'$ are as required. □

**Proposition 3.6.2.** *Let F be an irreducible covering of a pre-smooth set D, $\langle a, b \rangle \in F$, and assume that $a \in D$ is regular for F. Then for every $a' \in V_a \cap {}^*D$, there exists $b' \in V_b$, such that $\langle a', b' \rangle \in F$ and $\mathrm{cdim}\,(b'/a')$ is equal to r, the dimension of generic fibre of F.*

*Proof.* We first find *some* $b^0 \in V_b$ such that $\langle a', b^0 \rangle \in F$.

*Case (i).* Assume that $a' \in V_a$ is generic in $D({}^*M)$ (over $M$). Consider the type over ${}^*M$,

$$p(y) = \{F(a', y)\} \cup \mathrm{Nbd}_b(y).$$

Claim: $p$ is consistent.

Proof of claim: Assume not. Then $\forall y (F(a', y) \to Q(c', y))$ holds for some $Q(z, y)$ and $c'$ is as in the definition of $\mathrm{Nbd}_b$. Let $L(z, y)$, $\hat{L}(z, y)$, and $D'$ be as in Proposition 3.6.1. Because $a'$ is generic in $D$ and $L(c', a')$ holds, we have $\hat{L}(c,' a')$ and hence $\hat{L}(c, a)$. Because $a$ is regular, we have $L(c, a)$. But $F(a, b)$ holds, so by the definition of $L$, we get $Q(c, b)$, contradicting the choice of $Q$. Claim proved.

By Lemma 2.2.21, the consistency of $p$ implies the existence of $b^0 \in V_b$ such that $\models F(a', b^0)$. Case (i) solved.

*Case (ii).* Let now $a' \in V_a$ be arbitrary. We want to find $b^0 \in V_b$ such that $\models F(a', b^0)$.

Let ${}^{**}M \succeq {}^*M$ and $\pi^* : {}^{**}M \to {}^*M$ be as in Lemma 2.2.17. (Notice that ${}^*M$ is a Zariski structure but not necessarily pre-smooth, and similarly we do not know whether ${}^*D$ is pre-smooth with ${}^*M$.) Let ${}^*V_a = (\pi \circ \pi^*)^{-1}(a)$, and for $a' \in {}^*M$ let ${}^{**}V_{a'} = (\pi^*)^{-1}(a')$.

By Proposition 3.5.28, there is $a'' \in {}^{**}V_{a'}$, $a''$ generic in $D$ over ${}^*M$. But then $a''$ is generic over $M$ hence, by case (i), there is $b'' \in {}^*V_b$ such that $(a'', b'') \in F$. We have then $F(\pi^*(a''), \pi^*(b''))$ where $\pi^*(a'') = a'$ and $b^0 = \pi^*(b'')$ is in $V_b$.

Now we want to replace $b^0$ by a generic element. Let $F^0(y)$ be the (${}^*M$-definable) connected component of $F(a', y)$ containing $b^0$. By Lemma 3.5.30, $\dim F^0(y) = r$. Choose $b'$ to be generic in $F^0$, over $M \cup \{a'\}$, then $\mathrm{cdim}\,(b'/{}^*M) = r$. For every $Q$ as in $\mathrm{Nbd}_b$, we must have $\dim(Q(c', {}^*M) \cap$

$F^0(a', {}^*M)) < r$ because otherwise $\neg Q(c', {}^*M) \cap F^0(a', {}^*M) = \emptyset$, contradicting the existence of $b^0$. Clearly then, $b' \models \mathrm{Nbd}_b$; thus, by the existential closedness of ${}^*M$ and Lemma 2.2.21, we can choose $b' \in \mathcal{V}_b$ with $\mathrm{cdim}\,(b'/a') = r$. $\qquad\square$

**Lemma 3.6.3.** *Let $F$ be an irreducible covering of a pre-smooth $D$, $a \in D$, $a' \in \mathcal{V}_a \cap D$ generic, and $b^0 \in \mathcal{V}_b$, $\langle a', b^0 \rangle \in F$.*
*Then there is $b' \in \mathcal{V}_b$ such that $\models F(a', b')$ and $\mathrm{cdim}\,(b'/a')$ is equal to $r$, the dimension of generic fibre of $F$.*

*Proof.* See the last part of the proof of Proposition 3.6.2. $\qquad\square$

**Example 3.6.4.** We show here that it is essential in Proposition 3.6.2 that $a$ is regular.

Let $\mathrm{M} = \mathbb{C}$, the field of complex numbers, $D = \mathbb{C} \times \mathbb{C}$ and $F(x_1, x_2, y)$ be given by the formula $x_1 \cdot y = x_2$. Then the dimension of generic fibre of $F$ is 0 while the dimension of the fibre $F(0, 0, M)$ is 1. So, $(0, 0)$ is not regular. Choose distinct $b, c \in \mathbb{C}$ and $a' \in \mathcal{V}_0$, $a' \neq 0$. Then $F(0, 0, b)$ holds and $(a', a'c) \in \mathcal{V}_{(0,0)}$. Obviously, the only possible value for $y$ in $a' \cdot y = a'c$ is $c$ and $c \notin \mathcal{V}_b$, so Proposition 3.6.2 fails in the point $(0, 0)$.

**Example 3.6.5.**
1. Consider the plane complex curve $F$ given by the equation

$$x - y^3 - y^2 = 0.$$

The projection $\langle x, y \rangle \mapsto x$ is a covering of $\mathbb{C}$. Proposition 3.6.2 tells that for any $\alpha \in \mathcal{V}_0$, there is $\beta \in \mathcal{V}_0$, such that $\langle \alpha, \beta \rangle \in F$. In fact, the $\beta$ in the neighbourhood is defined uniquely because the equation $\alpha = y^3 + y^2$ has only one solution in the infinitesimal neighbourhood of 0. Notice that we cannot use the classical implicit function theorem here, because $\partial f / \partial y = 0$ at 0 for $f(x, y) = x - y^3 - y^2$.

2. Consider the plane curve $F$ given by the equation $x^2 - y^p = 0$ in an algebraically closed field of characteristic $p$. Again we can solve the equation locally, $y = \varphi(x)$ near 0, with $\varphi$ a 'local Zariski continuous function', though the implicit function theorem is not applicable.

## 3.6.2 Multiplicities

Recall the definition of a finite covering from Definition 3.5.15.

**Lemma 3.6.6.** *Let $F \subseteq D \times M^k$ be an irreducible finite covering of $D$ in $a$, $D$ being pre-smooth. If $F(a, b)$ and $a' \in V_a \cap D(*M)$ is generic in $D$, then*

$$\#(F(a', {}^*M) \cap V_b) \geq \#(F(a'', {}^*M) \cap V_b), \qquad \text{for all } a'' \in V_a \cap D(*M)$$

*Proof.* Take $a'' \in V_a \cap D(*M)$ and assume that $b_1'', \ldots, b_m'' \in V_b$ are distinct and that $F(a'', b_1''), \ldots, F(a'', b_m'')$ hold. Let $F_0^{(m)}$ be an irreducible component of the set defined by $F(x, y_1) \& \cdots \& F(x, y_m)$ which contains $\langle a'', b_1'', \ldots, b_m'' \rangle$. Notice that $\dim F = \dim D$; hence, by Exercise 3.5.7(2), $\dim F_0^{(m)} = \dim D$. The projection of $F_0^{(m)}$ into $D$ has finite fibre at $a$. It follows by (AF) that $F_0^{(m)}$ is a cover of $D$ and the point $a$ is regular for $F_0^{(m)}$.

By applying $\pi$ to $\langle a'', b_1'', \ldots, b_m'' \rangle$, we get $F^{(m)}(a, b, \ldots, b)$, and by Proposition 3.6.2, because $a' \in V_a$, there are $b_1', \ldots, b_m' \in V_b$ such that $F_0^{(m)}(a', b_1', \ldots, b_m')$.

Consider the open set

$$U_m = \left\{ \langle x, y_1, \ldots, y_m \rangle : \bigwedge_{i \neq j} y_i \neq y_j \right\}.$$

By our assumption, $U_m \cap F^{(m)} \neq \emptyset$; hence, every generic point of $F^{(m)}$ lies in $U_m$. However, $\langle a', b_1', \ldots, b_m' \rangle$ is generic in $F^{(m)}$, and hence $\bigwedge_{i \neq j} b_i' \neq b_j'$. $\square$

**Definition 3.6.7.** Let $\langle a, b \rangle \in F$ and $F$ be a finite covering of $D$ in $\langle a, b \rangle$. Define

$$\text{mult}_b(a, F/D) = \#F(a', {}^*M^k) \cap V_b, \qquad \text{for } a' \in V_a \text{ generic in } D \text{ over } M.$$

By Lemma 3.6.6, this is a well-defined notion, independent of the choice of generic $a'$. Moreover, the proof of Lemma 3.6.6 also contains the proof of the following.

**Lemma 3.6.8.** $m \leq \text{mult}_b(a, F/D)$ *iff there is an irreducible component $F_0^{(m)}$ of the covering $F(x, y_1) \& \cdots \& F(x, y_m)$ of $D$, finite at $a$, such that for any generic $a' \in V_a \cap D(*M)$ there are distinct $b_1', \ldots, b_m' \in V_b$ with $\langle a', b_1', \ldots, b_m' \rangle \in F_0^{(m)}$.*

Call a finite covering **unramified** at $\langle a, b \rangle$ if $\text{mult}_b(a, F/D) = 1$ and let

$$\text{unr}(F/D) = \{ \langle a, b \rangle \in F : \text{mult}_b(a, F/D) = 1 \}.$$

Assuming $a \in \text{reg}(F/D)$, set

$$\text{mult}(a, F/D) = \sum_{b \in F(a, M^k)} \text{mult}_b(a, F/D).$$

**Proposition 3.6.9 (multiplicity properties).** *Suppose D is pre-smooth. Then*

*(i) the previous definitions do not depend on the choice of* $^*M$ *and* $\pi$ *;*

*(ii)*

$$\text{mult}(a, F/D) = \#F(a', {}^*M^k)$$

*for* $a' \in D(^*M)$ *generic over M (not necessarily in* $V_a$ *), and the number does not depend on the choice of a in D;*

*(iii) the set*

$$j_m(F/D) = \{\langle a, b \rangle : a \in \text{reg}(F/D) \,\&\, \text{mult}_b(a, F/D) \geq m\}$$

*is definable and relatively closed in the set* $\text{reg}(F/D) \times M^k$ *. Moreover, there is m such that for every* $a \in \text{reg}(F/D)$ *, we have* $\text{mult}_b(a, F/D) \leq m$ *.*

*(iv)* $\text{unr}(F/D)$ *is open in F and the set*

$$D_1 = \{a \in \text{reg}\,(F/D) : \forall b\,(F(a, b) \to \langle a, b \rangle \in \text{unr}\,(F/D))\}$$

*is dense in D.*

**Remark.** In classical algebra-geometric context, $j_m(F/D)$ is defined in terms of the length of the corresponding localisation of a certain commutative coordinate ring. (See De Piro, 2004, for a comparative study of the notions of multiplicity.)

*Proof.*

(i) Assume that $\langle M', \pi' \rangle$ is another universal pair and that $\text{mult}_b(a, F/D) \geq m$, calculated with respect to $M'$. This implies the consistency of a certain type, which by the universality of $\langle M^*, \pi \rangle$ must also be realized in $^*M$, thus implying that $\text{mult}_b(a, F/D) \geq k$, calculated with respect to $^*M$.

(ii) is immediate from the definitions.

(iii) Given $m$, consider the set $F^{(m)}(x, y_1, \ldots, y_m) = F(x, y_1)\,\&\,\cdots\,\&\,F(x, y_m)$.

Let

$$U_m = \left\{ \langle x, y_1, \ldots, y_m \rangle : \bigwedge_{i \neq j} y_i \neq y_j \right\},$$

and let $F_1^{(m)}, \ldots, F_l^{(m)}$ be all those components of $F^{(m)}$ which have a non-empty intersection with $U_m$. By Lemma 3.6.8,

$$j_m\,(F/D) = \left\{ \langle a, b \rangle : a \in \text{reg}\,(F/D),\, M \models \bigvee_{i=1}^{l} F_i^{(m)}(a, b, \ldots, b) \right\}.$$

(The set on the right is relatively closed in $\text{reg}(F/D) \times M^k$.) By (DCC), there is an $m$ such that for every $p \geq m$ we have $j_m(F/D) = j_p(F/D)$.

(iv) First notice that by (iii), $\text{unr}(F/D)$ is open in $F$. It follows that $\dim F \backslash \text{unr}(F/D) < \dim D$, but $D_1$ can be obtained as the complement of $\text{pr}(F \backslash \text{unr}(F/D))$; hence, $D_1$ is dense in $D$.                               □

**Example 3.6.10.** Assume now that $M = K$, when $K$ is an algebraically closed field, as in Example 3.5.10. Consider the two curves discussed in the example. $C_1 = \{y^2 = x^3\}$ and $C_2 = \{y^2 = x^2 + x^3\}$ can be considered coverings of the affine line $K$, $(x, y) \mapsto x$.

$\#C_1(a, K) = 2$ at every $a \in K$ except $a = 0$, where $\#C_1(a, K) = 1$. It follows that $C_1$ is an unramified cover of $K$ at every point except $(0, 0)$, and we have

$$\text{mult}_0(0, C_1/K) = 2.$$

We would get the same situation, with multiplicity 3, if we consider $C_1$ as a cover $(x, y) \mapsto y$ of $K$.

$C_2$ as a cover $(x, y) \mapsto x$ of $K$ behaves very similarly to $C_1$. It is an unramified cover of $K$ at every point except 0, where it has multiplicity 2.

If we consider $C_2$ as a cover $(x, y) \mapsto y$ of $K$, then $\#C_2(K, a) = 3$ for every $a$ except $a = 0$, $a = \pm\sqrt{4/27}$. For these last three points, we have $\#C_2(K, a) = 2$. The cover is unramified at every $(b, a)$ where $a \neq 0, \pm\sqrt{4/27}$. It is also unramified at the points $(-1, 0)$, $(2/3, \pm\sqrt{4/27})$ assuming the characteristic of $K$ is not 3. Indeed, let us see what happens e.g. at $(-1, 0)$. For $\alpha \in V_0$, we look for solutions of $\alpha^2 = x^2 + x^3$ near $-1$ (i.e. of the form $x = -1 + \beta$, $\beta \in V_0$). We thus have

$$\alpha^2 = (-1 + \beta)^2 + (-1 + \beta)^3 = \beta - 2\beta^2 + \beta^3.$$

So, if for the given $\alpha$ there is another solution $\beta' \in V_0$, then

$$0 = (\beta - \beta')(1 - 2(\beta + \beta') + \beta^2 + \beta\beta' + \beta'^2)$$

and only $\beta = \beta'$ is possible.

At the points $(0, 0)$ and $(-2/3, \pm\sqrt{4/27})$, the covering has multiplicity 2.

**Exercise 3.6.11.** *Let $F \subseteq D \times M^k$ be a finite covering of $D$ where $D$ is presmooth. Let $D_1$ be the set from Proposition 3.6.9(iv), and let $s = \#F(a', {}^*M)$ for $a'$ generic in $D$. Then*

$$D_1 = \{a \in \text{reg}(F/D) : \#F(a, M) = s\}.$$

### 3.6.3 Elements of intersection theory

**Definition 3.6.12.** Let $P$ and $L$ be constructible irreducible sets and $I \subseteq P \times L$ be closed in $P \times L$ and irreducible, $\mathrm{pr}_2 I = L$. We call such an $I$ a **family of closed subsets of** $P$. One can think of $l \in L$ as the parameter for a closed subset $\{p \in P : pIl\}$.

Any $l \in L$ identifies a subset of those points of $P$, that are incident to $l$, though we allow two distinct $l$'s of $L$ to represent the same set. As a rule, we write simply $p \in l$ instead of $pIl$. Thus, the mentioning of $I$ is omitted, and we simply refer to $L$ as a family of closed subsets of $P$.

**Definition 3.6.13.** Let $L_1$ and $L_2$ be irreducible families of closed subsets of an irreducible set $P$. We say that the **families intersect in a finite way** if for any generic pair $\langle l_1, l_2 \rangle \in L_1 \times L_2$ the intersection $l_1 \cap l_2$ is non-empty and finite. In this situation, for $p \in P$ and $l_1 \in L_1, l_2 \in L_2$ such that $l_1 \cap l_2$ is finite, define the **index of intersection of** $l_1, l_2$ **at the point** $p$ **with respect to** $L_1, L_2$ as

$$\mathrm{ind}_p\,(l_1, l_2/L_1, L_2) = \#l_1' \cap l_2' \cap V_p,$$

where $\langle l_1', l_2' \rangle \in V_{l_1, l_2} \cap {}^*L_1 \times {}^*L_2$ is generic over $M$.

**Definition 3.6.14.** The index of intersection of these two families is

$$\mathrm{ind}(L_1, L_2) = \#l_1' \cap l_2'$$

where $\langle l_1', l_2' \rangle \in {}^*L_1 \times {}^*L_2$ is generic over $M$.

**Proposition 3.6.15.** *Assume that $M$ is complete. Assume also that $L_1 \times L_2$ and $P \times L_1 \times L_2$ are pre-smooth and irreducible and that the families intersect in a finite way. Then*

*(i) the definition of the index at a point does not depend on the choice of* ${}^*M$, *$\pi$, and generic $l_1', l_2'$;*

*(ii)*

$$\sum_{p \in l_1 \cap l_2} \mathrm{ind}_p\,(l_1, l_2/L_1, L_2) = \mathrm{ind}(L_1, L_2);$$

*(iii) for generic $\langle l_1, l_2 \rangle \in L_1 \times L_2$ and $p \in l_1 \cap l_2$*

$$\mathrm{ind}_p\,(l_1, l_2/L_1, L_2) = 1;$$

*(iv) the set*

$$\{\langle p, l_1, l_2 \rangle \in P \times L_1 \times L_2 : \mathrm{ind}_p (l_1, l_2/L_1, L_2) \geq k\}$$

*is closed.*

*Proof.* This is contained in the properties of multiplicities for finite coverings. Let

$$D = \{\langle l_1, l_2 \rangle \in L_1 \times L_2 : l_1 \cap l_2 \text{ is non-empty and finite}\}.$$

This is an open subset of $L_1 \times L_2$. Let

$$F = \{\langle p, l_1, l_2 \rangle : \langle l_1, l_2 \rangle \in D, \ p \in l_1 \cap l_2\}.$$

This is a covering (maybe reducible) of $D$. To apply Proposition 3.6.9, notice that by pre-smoothness any component $F_i$ of $F$ is of dimension $\dim D$; hence, the projection $\mathrm{pr}\, F_i$ of $F_i$ on $D$ is dense in $D$ and $F_i$ is finite in $\langle p, l_1, l_2 \rangle$. By completeness, $\mathrm{pr}\, F_i = D$. So, for each $F_i$ we may apply Proposition 3.6.9. Obviously,

$$\mathrm{ind}_p (l_1, l_2/L_1, L_2) = \sum_i \mathrm{mult}_p (\langle l_1, l_2 \rangle, F_i/D)$$

and the statements of the proposition follow. □

**Remark 3.6.16.** The proposition effectively states that closed subsets from a given pre-smooth family are numerically equivalent (Hartshorne 1977).

**Exercise 3.6.17 (Problem).** *Develop a theory of intersection and of numerical equivalence of closed sets in pre-smooth Zariski structures.*

**Definition 3.6.18.** Suppose for some $\langle l_1, l_2 \rangle \in L_1 \times L_2$ that $l_1 \cap l_2$ is finite. Two closed sets $l_1$, $l_2$ from families $L_1, L_2$, respectively are called **simply tangent at the point** $p$ **with respect to** $L_1, L_2$ if there is an infinite irreducible component of $l_1 \cap l_2$, containing $p$ or

$$\mathrm{ind}_p (l_1, l_2/L_1, L_2) \geq 2.$$

We study the tangency in projective spaces in Section 4.3 and also a more specific form of tangency between branches of curves at a fixed point in Section 3.8.

### 3.6.4 Local isomorphisms

**Definition 3.6.19.**

(i) Let $F \subseteq D \times M^k$ be a definable relation, $\langle a, b \rangle \in F$. We say that $F$ **defines a local function from** $\mathcal{V}_a \cap D$ **into** $\mathcal{V}_b$ if $F|(\mathcal{V}_a \times \mathcal{V}_b)$ is the graph of a function from $\mathcal{V}_a \cap D$ into $\mathcal{V}_b$.

(ii) Let $F \subseteq D \times R$ be a finite-to-finite irreducible relation, relatively closed in $D \times R$, $\mathrm{pr}_D(F) = D$.

We say that $F$ **defines a local function on** $D$ [and if $\mathrm{pr}_R(F) = R$, a **local isomorphism between** $D$ **and** $R$] if for every $\langle a, b \rangle \in F$, $F$ defines a local function from $\mathcal{V}_a$ into $\mathcal{V}_b$ ($F|\mathcal{V}_a \times \mathcal{V}_b$ is the graph of a bijection between $\mathcal{V}_a \cap D$ and $\mathcal{V}_b \cap R$).

The following corollary is an immediate consequence of the definitions and of Proposition 3.6.2.

**Corollary 3.6.20.** *Let $F \subseteq D \times M^k$ be, generically, a finite covering of $D$, $D$ pre-smooth. Then $F$ is unramified at a point $(a, b) \in F$ iff $F$ defines a local function from $\mathcal{V}_a$ into $\mathcal{V}_b$. In particular, if $D_1 = \{a \in \mathrm{reg}(F/D) : \forall b(F(a, b) \rightarrow (a, b) \in \mathrm{unr}(F/D))\}$, then $F$ defines a local function on $D_1$.*

Finally, we omit in this corollary the assumption that $D$ is pre-smooth. We can do so if we work in a one-dimensional structure.

**Theorem 3.6.21.** *A one-dimensional, uncountable, pre-smooth, irreducible Zariski structure* M *is a Zariski geometry.*

First notice that M satisfies (EU). Indeed, one-dimensionality and irreducibility imply that any definable subset of $M$ is either finite or a complement to a finite set. Under the assumption of uncountability, it is easy to deduce from this that M is strongly minimal and indeed $\omega_1$-compact; thus, (EU) follows.

Now we are going to prove (sPS). We first prove the following lemma.

**Lemma 3.6.22.** *Assume that $F \subseteq D \times M^r$ is an irreducible cover of $D$ and $F$ defines a local function on $D$. If $D$ is pre-smooth, then so is $F$.*

*Proof.* Let $F'$ be a set of the form $F^k \times M^m$, $D' = D^k \times M^m$. By reordering the variables, we may consider $F'$ as a subset of $D' \times M^r$. It is then a finite cover of $D'$ and defines a local function on it.

Let $S_1, S_2$ be closed irreducible subsets of $F'$ and $\langle a, b \rangle \in D' \times M^r$, a point in $S_1 \cap S_2$. By Corollary 3.5.6, we just need to show that

$$\dim_{\langle a,b \rangle}(S_1 \cap S_2) \geq \dim S_1 + \dim S_2 - \dim F'.$$

Consider the point $(a, b, a, b)$ in the set $S_1 \times S_2 \cap \Delta$, where

$$\Delta = \{(x_1, y_1, x_2, y_2) \in D' \times M^r \times D' \times M^r : x_1 = x_2\}.$$

Because $D$ is pre-smooth and $S_1 \times S_2$, $\Delta$ are closed and irreducible, every component $K$ of $S_1 \times S_2 \cap \Delta$ satisfies

$$\begin{aligned}
\dim K &\geq \dim S_1 + \dim S_2 + \dim \Delta - 2\dim(D' \times M^r) \\
&= \dim S_1 + \dim S_2 + \dim D' + 2\dim M^r - 2\dim(D' \times M^r) \\
&= \dim S_1 + \dim S_2 - \dim D'.
\end{aligned}$$

Choose $K$, a component containing $(a, b, a, b)$, and let $(a_1, b_1, a_1, b_2) \in V_{(a,b,a,b)}$ be a generic element in $K$. Because $(a_1, b_1), (a_1, b_2)$ are in $V_{(a,b)} \cap F'$, and because $F'$ defines a local function on $D'$, we must have $b_1 = b_2$. That is, $(a_1, b_1)$ is in $S_1 \cap S_2$ and $\mathrm{cdim}(a_1, b_1/M) \geq \dim S_1 + \dim S_2 - \dim D'$. Because $F'$ is a finite cover of $D'$, we have $\dim D' = \dim F'$; hence, we showed

$$\dim_{(a,b)}(S_1 \cap S_2) \geq \dim S_1 + \dim S_2 - \dim F'.$$

$\square$

We can now prove Theorem 3.6.21: By our assumption, there is a projection $\mathrm{pr} : M^m \to M^n$ such that $S$ is, generically, a finite cover of $M^n$ [see Exercise 3.2.2(1)]. Because $M^n$ is pre-smooth, we can use Corollary 3.6.20 to obtain an open dense $D \subseteq M^n$ such that $S_1 = S \cap (D \times M^{m-n})$ defines a local function on $D$. By Exercise 3.5.7, $D$ is pre-smooth, so we can apply Lemma 3.6.22 to conclude that $S_1$ is pre-smooth. $\square$

**Theorem 3.6.23 (implicit function theorem).** *Let M be strongly pre-smooth Zariski structure (e.g. one-dimensional, pre-smooth), $D \subseteq M^n$ be irreducible, and $F \subseteq D \times M^r$ be an irreducible finite covering of $D$, $\dim F = \dim D$. Then there is an open dense subset $D_1 \subseteq D$ such that $F \cap (D_1 \times M^r)$ defines a local function on $D_1$.*

*Proof.* Without loss of generality, $F$ is a finite covering of $D$. By strong pre-smoothness, there is an open dense subset $D_1 \subseteq D$ which is pre-smooth. Now apply Corollary 3.6.20. $\square$

Our next goal is to show that a local isomorphism preserves the pre-smoothness property between sets.

**Lemma 3.6.24.** *Let $D, R$ be irreducible sets and assume that $F \subseteq D \times R$ defines a local function on $R$. If $T \subseteq R$ is irreducible, then any component of $Q = \{x \in D : \exists y \in T(y) \ \& \ F(x, y)\}$ is of the same dimension as $T$.*

*Proof.* For $a \in Q$, let $b \in T$ be such that $F(a, b)$. Because $F$ defines a local function on $R$, given $b' \in V_b \cap T(*M)$, $b'$ generic in $T$, there is $a' \in V_a \cap D(*M)$ such that $F(a', b')$; hence, $a' \in Q(*M)$. By Exercise 3.5.32(5), $\text{cdim}(a'/M) = \text{cdim}(b'/M)$. Hence

$$\max\{\text{cdim}(a'/M) : a' \in V_a \cap Q\} \geq \max\{\text{cdim}(b'/M) : b' \in V_b \cap T\}.$$

By Corollary 3.5.29, $\dim_a Q \geq \dim T$, and because $F$ is a finite cover, we have $\dim_a Q = \dim T$. It is easy to see that for every component $K$ of $Q$ there is $a \in Q$ such that $\dim K = \dim_a Q$; hence, $\dim K = \dim T$. $\qquad \square$

**Lemma 3.6.25.** *Let $D, R$ be irreducible sets and $F \subseteq D \times R$ be a local function on $R$. Assume further that $F$ is the graph of a continuous function $p : D \to R$. If $D$ is pre-smooth, then so is $R$.*

*Proof.* Let $R' = R^k \times M^m$, $D' = D^k \times M^m$, and $p' : D' \to R'$ be a mapping which is $p$ on the first $k$ coordinates and the identity on the rest $m$. The graph of $p'$ is a local isomorphism between $D'$ and $R'$. Take $T_1, T_2 \subseteq R'$ irreducible and $t \in T_1 \cap T_2$ generic in a component $T$ of $T_1 \cap T_2$. Then there is $q \in Q_1 \cap Q_2$ such that $p'(q) = t$ and $Q_1$, $Q_2$ are connected components containing $q$ of $p'^{-1}(T_1)$, $p'^{-1}(T_2)$, respectively. Let $Q$ be a component of $Q_1 \cap Q_2$ containing $q$. Then $p'$ is continuous. Hence, $p'(Q)$ is irreducible, and because $T \subseteq p'(Q) \subseteq T_1 \cap T_2$, we must have $p'(Q) = T$. By the pre-smoothness of $D$, we have

$$\dim Q \geq \dim Q_1 + \dim Q_2 - \dim F'.$$

By the previous lemma, the right-hand side of the equation equals $\dim T_1 + \dim T_2 - \dim R'$ and, because $Q$ is a component of $P'^{-1}(T)$, $\dim T = \dim Q$. $\qquad \square$

**Proposition 3.6.26.** *Let $D \subseteq M^n$, $R \subseteq M^r$ be irreducible and locally isomorphic via $F \subseteq D \times R$. Assume further that $F$ is closed in $D \times M^r$. If $D$ is pre-smooth, then so is $R$.*

*Proof.* $F$ is a local function on $D$; hence, by Lemma 3.6.22, $F$ is pre-smooth. The graph of the projection map $\text{pr}_R : F \to R$ is easily seen to define a local function on $R$; thus by Lemma 3.6.25 (with $F$ in the role of $D$ now), $R$ is pre-smooth. $\qquad \square$

Even though Theorem 3.6.23 contains the assumption (used in the proof) that $F$ is relatively closed in $D \times M^k$, we can now do away with it.

**Exercise 3.6.27.** *Prove the following modification of Theorem 3.6.23: Let* M *be pre-smooth, one-dimensional, irreducible Zariski structure, let $D \subseteq M^n$ be irreducible, and let $F \subseteq D \times M^r$, $\dim F = \dim D$, be an irreducible set whose projection on D is surjective.*

*Then there is an open dense subset $D_1 \subseteq D$ such that $F \cap (D_1 \times M^r)$ is relatively closed and defines a local function on $D_1$.*

**Exercise 3.6.28.** *Let* M *be pre-smooth, one-dimensional, irreducible Zariski structure, $D \subseteq M^m$ and $R \subseteq M^r$. Let $F \subseteq D \times R$ be irreducible, $\mathrm{pr}_D(F) = D$, $\mathrm{pr}_R(F) = R$, and F a finite-to-finite relation. Then there are $D_1$ and $R_1$, open and dense in D and R, respectively, such that F defines a local isomorphism between $D_1$ and $R_1$. (In particular, $F \cap D_1 \times R_1$ needs to be relatively closed.)*

**Definition 3.6.29.** Under the assumption that M is one-dimensional pre-smooth, we call definable $D \subseteq M^n$ **smooth** if $D$ is locally isomorphic to an open subset of $M^k$ for some $k$.

Notice that under this terminology M itself is smooth.

**Theorem 3.6.30 (smoothness theorem).** *Assuming* M *is one-dimensional pre-smooth,*

(i) *any open subset of $M^n$ is smooth;*
(ii) *for every irreducible definable $D \subseteq M^n$ there is an open irreducible $D^0 \subseteq$ D which is smooth; and*
(iii) *if $D_1$ and $D_2$ are smooth, then so is $D_1 \times D_2$.*

*Proof.* Part (i) is part of the definition. Part (ii) is in fact proved in Theorem 3.6.21, and Part (iii) is immediate from the definition.                    □

## 3.7 Getting new Zariski sets

Some constructions in later parts of the notes and, more generally, in algebraic geometry lead us to consider more complex definable sets (and structures) which can be seen as a Zariski geometry compatible with the initial structure. We discuss two of such constructions in this section.

We fix now a Zariski structure M and consider a constructible irreducible subset $N \subseteq M^n$ and a (relatively) closed equivalence relation $E$ on $N$. We take $p : N \to N/E$ to denote the canonical projection mapping and use $p$ also for the induced map from $N^k$ onto $(N/E)^k$. We equip $N/E, (N/E)^2, \ldots$ with a topology as follows:

**Definition 3.7.1.** A subset $T \subseteq (N/E)^k$ is called **closed** in $(N/E)^k$ if $p^{-1}(T)$ is closed in $N^k$. Sets of the form $(N/E)$ together with the structure of closed subsets will be called **topological sorts in** $M$.

Notice that we can identify $(N/E)^k$ with the quotient $N^k/E^{(k)}$, where

$$(a_1, \ldots, a_k)E^{(k)}(b_1, \ldots, b_k) \Leftrightarrow a_i E^{(k)} b_i, \qquad i = 1, \ldots, k.$$

The topology we put on $(N/E)^k$ is then exactly the quotient topology induced from $N^k$.

**Notation.** We use $E(a, b)$ and $aEb$ interchangeably. For $s \in N^k$, we denote by $sE$ the $E^{(k)}$-equivalence class of $s$.

**Lemma 3.7.2.** *Every topological sort satisfies (L), (DCC), and (SP) [or (P) if $M$ is complete].*

*Proof.* Immediate from definitions. □

**Lemma 3.7.3.**

(i) *The map $p : N \to N/E$ is a continuous, closed, and open map.*
(ii) *$T \subseteq (N/E)^n$ is irreducible iff there is an irreducible $S \subseteq N^n$ such that $p(S) = T$.*

*Proof.* (i) If $T = p(S)$, then

$$p^{-1}(T) = \{a \in N : (\exists b \in S) aEb\}.$$

So, if $S$ is closed, then so is $p^{-1}(T)$ and hence $T$ is closed. However, $p^{-1}(T)$ is the complement of the set $\{a \in N : \exists b \ (b \notin S \ \& \ aEb)\}$. Hence, if $S$ is open so is $T$.

Part (ii) follows easily from (i). □

**Definition 3.7.4.** For $N/E$, a topological sort, assume that $T = p(S) \subseteq (N/E)^k$ for some $S \subseteq N^k$ closed irreducible (hence $T$ also is). Define

$$\dim(T/S) = \dim(S) - \min\{\dim(p^{-1}(t) \cap S) : t \in T\}.$$

As we show, this definition does not depend on the choice of $S$. We first introduce an alternative way of defining dimension: for $S \subseteq (N/E)^k$ and $l \geq 0$, define $\delta_S^l = \dim\{a \in S : \dim(aE \cap S) = l\}$, and let

$$\delta_S = \max\{\delta_S^l - l : l \geq 0\}.$$

For $S$ irreducible and $T = p(S)$, if $l = \min\{\dim(p^{-1}(t) \cap S) : t \in T\}$, then the set $\{a \in S : \dim(aE \cap S) = l\}$ is open and dense in $S$; hence, $\delta_S^l = \dim S$ and so $\dim(T/S) = \delta_S$.

**Lemma 3.7.5.** *For any irreducible closed $S \subseteq N^k$, if $S_1 \subseteq N^k$ is closed and $p(S) = p(S_1)$, then*

*(i)* $\delta_{S_1} = \dim S_1 - m$ *for* $m = \min\{\dim(aE \cap S_1) : a \in S_1\}$.
*(ii)* $\delta_S = \delta_{S_1}$. *In particular, if $S_1$ is irreducible, then* $\dim(T/S) = \dim(T/S_1)$.

*Proof.* Let $m = \min\{\dim(aE \cap S_1) : a \in S_1\}$, and let $U$ be the set of all $a \in S$ such that $\dim(aE \cap S_1) = m$. By Lemma 3.7.3 [and the fact that $p(S) = p(S_1)$], $U$ is open and dense in $S$.

(i) If $D$ is an irreducible component of $S_1$, let $m(D) = \min\{\dim(aE \cap S_1) : a \in D\}$. The set

$$V(D) = \{a \in D : \dim(aE \cap S_1) = m(D)\}$$

is open in $D$ and hence in $S_1$. By Lemma 3.7.3, $p^{-1}p(V(D)) \cap S$ is open, non-empty, and hence dense in $S$, but then it must intersect $U$ so $m(D) = m$. We showed then that there is an open dense subset $V$ of $S_1$ such that if $a \in V$, then the intersection of $aE$ with $S_1$ has minimal dimension. By the definition of $\delta$, we get

$$\delta_{S_1} = \dim S_1 - m.$$

(ii) Let $l = \min\{\dim(aE \cap S) : a \in S\}$, and let $U \subseteq S$ be now all $a \in S$ such that $\dim(aE \cap S) = l$ and $\dim(aE \cap S_1) = m$. Again, $U$ is an open dense subset of $S$ and just like in part (i), we can show that $V = p^{-1}p(U)$ is an open dense subset of $S_1$. We take $\bar{E}_1 \subseteq S \times S_1$ to be the topological closure of $E_1 = E \cap (U \times V)$ and let $pr_1, pr_2$ be the projections on the first and second co-ordinates, respectively.

Take $K$ to be a component of $\bar{E}_1$ of maximal dimension. $K$ is the closure of a component of $E_1$; hence, $pr_1(K) \cap U \neq \emptyset$. It follows that $\min\{\dim K(a, S_1) : a \in pr_1(K)\} \leq m$, so $\dim K \leq \dim S + m$. However, $\dim K = \dim \bar{E}_1 \geq \dim S + m$ (see **Fact 3.1.7**), therefore $\dim K = \dim S + m$. Similarly (taking into account that $V$ may not be irreducible), $\dim K = \dim V + l = \dim S_1 + l$; hence

$$\dim S - l = \dim S_1 - m.$$

$\square$

The lemma allows us to define dim $T$ as $\dim(T/S)$ for any irreducible closed $S$ such that $p(S) = T$ independent of $S$.

**Definition 3.7.6.** For $N/E$, a topological sort, if $T \subseteq (N/E)^k$ is closed and $T = \bigcup_{i \leq k} T_i$ is (the unique) irreducible decomposition of $T$, define

$$\dim(T) = \max_{i \leq k} \dim(T_i).$$

It is easy to see that this definition agrees with $\delta_S$ for any closed set $S$ such that $p(S) = T$: Let $S = \bigcup_{i \leq k} S_i$, for $S_i$ closed irreducible, $p(S_i) = T_i$. We denote by $S_i'$ the set $p^{-1}p(S_i) \cap S$. Then $S = \bigcup_{i \leq k} S_i'$, and it easily follows from the definition that $\delta_S = \max\{\delta_{S_i'} : i \leq k\}$, which equals $\max\{\dim T_i : i \leq k\}$.

**Lemma 3.7.7.** *A topological sort with the notion of dimension as before satisfies (DU), (SI), and (DP).*

*Proof.* (DP) is immediate, and (DU) easily follows from (SI).

To prove (SI), let $T_1 \subset T_2$ be two irreducible closed sets in a closed topological sort, $T = N/E$. There are then $S_1 \subset S_2 \subseteq N^k$ irreducible closed such that $p(S_i) = T_i, i = 1, 2$. Let $S_1' = p^{-1}p(S_1) \cap S_2$. Then it is sufficient to show that $\delta_{S_1'} < \delta_{S_2}$. Because $S_1' \subset S_2$, we have, $\dim S_1' < \dim S_2$, and also

$$\min\{\dim(aE \cap S_1') : a \in S_1'\} \geq \min\{\dim(aE \cap S_2) : a \in S_2\}.$$

By Lemma 3.7.5(i), $\delta_{S_1'} < \delta_{S_2}$. $\qquad\square$

**Lemma 3.7.8.** *Any definable subset $R \subseteq (N/E)^n$ is a Boolean combination of closed subsets.*

*Proof.* Use the elimination of quantifiers in $M$ to see that $p^{-1}(R)$ is a Boolean combination of closed sets, and then use Lemma 3.7.3 to show that $R$ is. $\quad\square$

**Example 3.7.9.** Let $N = \mathbf{P}^1 \times \mathbf{P}^1$ ($\mathbf{P}^1$ is the projective line over an algebraically closed field), $a \in \mathbf{P}^1$, and $E$ be an equivalence relation on $N$ whose classes are either a copy of the $y$-axis or singletons not on that axis. Namely, for $\langle x, y \rangle, \langle x', y' \rangle \in N$

$$\langle x, y \rangle E \langle x', y' \rangle \quad \text{iff} \quad x = x' \,\&\, (y = y' \lor x = a).$$

Define $S \subseteq N \times N$ to be the set

$$\{(\langle x_1, y_1 \rangle, \langle x_1, y_2 \rangle)) : x_1, y_1, y_2 \in P^1\},$$

and let $T = p(S)$. Now, if $t \in T$ is in $\{(a, y) : y \in \mathbf{P}^1\}$, then $T(t, N/E)$ contains exactly one element and hence has dimension 0. However, for any other $t$, we have $\dim T(t, N/E) = 1$. In particular, the set $\mathcal{P}(T, 0)$ is not closed, and hence (FC) does not hold.

**Exercise 3.7.10.**

*(i) Show that (AF) does not hold for N and E as in the previous example.*

*(ii) Find an example where, in the notations of the proof of Lemma 3.7.5, $\dim(\bar{E}_1) < \dim E$.*

**Definition 3.7.11.** A topological sort in M satisfying all the axioms of a Zariski structure will be called a **Zariski set (Z-set)** in $M$.

Given two topological sorts $T_1 = N_1/E_1$ and $T_2 = N_2/E_2$, we can put a natural product structure on $T_1 \times T_2$, namely the one induced by the equivalence relation $E_1 \times E_2$ on $N_1 \times N_2$. For $S_1, S_2$ subsets of $T_1^k, T_2^l$, respectively, we call a map $\phi : S_1 \to S_2$ a **morphism** (Z-morphism) if the graph of $\phi$ is closed in $S_1 \times S_2$. If $\phi$ is a bijection, we say that $\phi$ is an **isomorphism** of $S_1$ and $S_2$.

**Exercise 3.7.12.** *For $S_1, S_2, T_1, T_2$ as before, if $\phi : S_1 \to S_2$ is a morphism, then an inverse image of a closed set under $\phi$ is closed (in the topology induced by $T_1, T_2$, respectively).*

**Definition 3.7.13.** A topological sort $T$ is called a **pre-manifold in M** if there exists a finite collection $U_1, \ldots, U_k$ of subsets which are open and dense in $T$ such that

(i) $T = U_1 \cup \cdots \cup U_k$; and
(ii) for every $i \leq k$ there is an irreducible subset $V_i \subseteq M^n$, pre-smooth with $M$, and an isomorphism $\phi_i : U_i \to V_i$.

**Remark 3.7.14.** For any $i, j$, if $U_i \cap U_j \neq \emptyset$, then the map $\phi_j \circ \phi_i^{-1}$ is an isomorphism between open subsets of $V_i$ and $V_j$.

**Definition 3.7.15.** A topological sort $T = N/E$ is called an **eq-fold** if $E$ is a **finite** equivalence relation, that is, with all classes finite.

**Proposition 3.7.16.** *Every pre-manifold $T$ is an irreducible pre-smooth Z-set.*

*Proof.* $T$ is irreducible. Indeed, every $U_i$ must be irreducible because $V_i$ are irreducible. If $T = S_1 \cup S_2$ for closed subsets $S_1$ and $S_2$, then by irreducibility $U_i \subseteq S_1$ or $U_i \subseteq S_2$. By the density of $U_i$, it correspondingly follows that $S_1 = T$ or $S_2 = T$. Irreducibility follows.

To prove the rest, after Lemmas 3.7.2 and 3.7.7, we need to check (AF), (FC), and (PS) only. All the three conditions are local; that is, it is enough to check the conditions for pr $S \cap U_i$ in the case of (AF) and (FC) for each $i$, and check it for $S_1 \cap U_i$ and $S_2 \cap U_i$ for (PS). This obviously holds because the conditions are preserved by continuous open dimension, preserving bijections $\phi_i$. $\qquad \square$

Our aim in the remaining part of this section is to prove an analogous result for eq-folds. We do that under some assumptions on $E$.

**Definition 3.7.17.** An equivalence relation $E$ on $N$ is called **e-irreducible** if for any irreducible component $E_i$ of $E$, both projections on $N$ are dense in $N$.

**Lemma 3.7.18.** *Let $\Gamma$ be a finite group acting on an irreducible $N$ by Zariski continuous bijections. Then the equivalence $E_\Gamma$, given by*

$$x E_\Gamma y \qquad \textit{iff} \qquad \exists \gamma \in \Gamma \ \gamma x = y,$$

*is e-irreducible.*

*Proof.* Obviously

$$E_\Gamma = \bigcup_{\gamma \in \Gamma} \text{graph } \gamma,$$

and each graph $\gamma$ is irreducible, isomorphic via projections to $N$. We have found the irreducible decomposition of $E_\Gamma$ which obviously satisfies the definition of e-irreducibility. $\qquad\square$

**Definition 3.7.19.** We call an eq-fold of the form $N/E_\Gamma$ an **orbifold**.

**Lemma 3.7.20.** *For any closed finite equivalence relation $E$ on an irreducible topological sort $N$,*

(i) *$\dim E = \dim N$, moreover $\dim E_i = \dim N$ for every component of $E$ which projects densely on $N$, and*
(ii) *there is an open dense $U \subseteq N$ such that $E \cap U^2$ is e-irreducible.*

*Proof.*

(i) $\dim E \geq \dim N$ because $E$ contains the diagonal. Then $\dim E \leq \dim N$ because the projection $E \to N$ has finite fibres.
(ii) Components $E_i$ of $E$ with small projections can be characterised by dimension $\dim E_i < \dim N$. For every such $E_i$, throw out the closure of the small pr $E_i$. The remaining $U$ satisfies the requirements. $\qquad\square$

**Lemma 3.7.21.** *Let $N/E$ be a topological sort, $E$ be a finite, closed, e-irreducible equivalence relation on $N$, and $N$ be pre-smooth with $M$. Let $T \subseteq (N/E)^k \times M^m$ be closed and irreducible. Then, every component of $p^{-1}(T) \subseteq N^k \times M^m$ is of dimension equal to $\dim T$.*

*Proof.* By an obvious isomorphism, we may assume that

$$T \subseteq (N^k \times M^m)/\tilde{E}$$

for $\tilde{E}$ on $N^k \times M^m$ defined as $E$ on the first $k$ co-ordinates and as equality on the last $m$ co-ordinates.

Now, if $S_0, S_1 \subseteq N^k \times M^m$ are components of $p^{-1}(T)$ and $\dim S_0 = \dim T$, then $p(S_0) = T \supseteq p(S_1)$. The latter means that each point of $S_1$ is $\tilde{E}$-equivalent to a point in $S_0$.

Let $\mathrm{pr}_i : S_0 \times S_1 \to S_i$, $i = 0, 1$, be the projections maps. Choose $\langle s_0, s_1 \rangle \in \tilde{E}$ with $s_1$ generic in $S_1$, and let $\tilde{E}_j$ be a component of $\tilde{E}$ containing $\langle s_0, s_1 \rangle$. Then $\tilde{E}_j \cap (S_0 \times S_1) \neq \emptyset$, and $\mathrm{pr}_1 \tilde{E}_j$ is dense in $S_1$.

Consider the set

$$\{\langle x, y \rangle : x, y \in N^k \times M^m \,\&\, x \in S_0 \,\&\, \langle x, y \rangle \in \tilde{E}_j\},$$

which can also be seen as the intersection of two subsets, $S_0 \times (N^k \times M^m)$ and $\tilde{E}_j$ of $(N^k \times M^m)^2$. By pre-smoothness, the dimension of any component $K$ of the set is not less than

$$(\dim S_0 + k \dim N + m \dim M) + (k \dim N + m \dim M)$$
$$-2(k \dim N + m \dim M) = \dim S_0.$$

On the other hand, such a $K$ projects into $S_0$ with finite fibres; hence, $\dim K = \dim S_0$ and $\mathrm{pr}_1(K)$ is dense in $S_0$.

Pick a $K$ containing this pair $\langle s_0, s_1 \rangle$. Then, $\mathrm{pr}_2(K)$ is dense in $S_1$. By the finiteness of $\tilde{E}$, we have $\dim S_1 = \dim K = \dim S_0$.

So, we have proven that all components of $p^{-1}(T)$ are of the same dimension equal to $\dim T$.                                                                    $\square$

**Proposition 3.7.22.** *Let $U = N/E$ be an eq-fold, $E$ be a finite, closed, e-irreducible equivalence relation on $N$, and $N$ be pre-smooth with $M$. Then $U$ is a Z-set pre-smooth with $M$. In particular, any orbifold on a pre-smooth $N$ is a pre-smooth Z-set.*

*Proof.* As in the proof of Proposition 3.7.16, we need to check only three conditions: (AF), (FC), and (PS). Now we notice that all three conditions are formulated in terms of dimensions and irreducible sets. Lemma 3.7.21 allows one to transfer the conditions from irreducible subsets of Cartesian products of $U$ and $M$ to those of $N$ and $M$, for which the properties hold by assumption.                                                                    $\square$

**Definition 3.7.23.** Let M be a Zariski structure (not necessarily pre-smooth) and $C$ an irreducible Z-set in M. We say that $C$ is **pre-smooth** if $C$ as a Zariski structure satisfies (PS).

We can now define the following useful notion.

**Definition 3.7.24.** Let M be a Zariski structure and $C$ an irreducible pre-smooth Zariski structure of dimension 1 (Zariski curve). Let $U$ be a Z-set in M. We call $U$ an $n$-**manifold with respect to** $C$ if $U$ is locally isomorphic to $C^n$, for some $n$.

## 3.8 Curves and their branches

In this section we develop, under certain technical assumptions, a theory of tangency between curves. This theory makes sense when we assume that the curves are members of 'nice' families. The theory of tangency is much easier in non-singular points, but we cannot assume that it is always the case. To deal with the more general situation, we introduce the notion of a branch of a curve at a point. We conclude the section with the proof that the tangency for branches is an equivalence relation. Moreover, this relation is definable.

We assume for the rest of this chapter and the next that M is a one-dimensional, irreducible, pre-smooth Zariski structure satisfying (EU) on the universe $C$. As we showed in Sub-section 3.5.3, any elementary extension of M is a pre-smooth Zariski structure. We work in a suitable elementary extension *M of M which is $\kappa$-saturated for a suitable $\kappa$ (the universal domain; see Section A.4.2). In particular, every definable set in *M contains generic points.

We also recall that by Theorem 3.2.8 such an *M model-theoretically is a strongly minimal structure with Morley rank rk equal to dimension dim. Thus, the usual dimension calculus (see Subsection B.1.2 [see Appendix] and Exercise 3.5.33) holds in *M, which we use later in several occasions.

**Definition 3.8.1.** By a **(definable) family of curves** in $C^m$, ($m \geq 2$) we mean a triple $(P, L, I)$ where $P$ is an open subset of $C^m$, $L$ is a $k$-manifold with respect to $C^k$, some $k \geq 1$ (Definition 3.7.24), $I \subseteq P \times L$ is an irreducible relation, closed in $P \times L$, and

(i) the corresponding projections of $I$ cover $P$ and $L$;
(ii) for every $l \in L$, the set $I(P, l)$ is one-dimensional, and for $l$ generic the set is irreducible;
   We call $I$ an **incidence relation for family** $L$.
   We say that the family is **faithful** if also

(iii) for any $l_1, l_2 \in L$, the intersection $I(P, l_1) \cap I(P, l_2)$ is finite (or empty) provided $\mathrm{acl}(l_1) \neq \mathrm{acl}(l_2)$.

We say that $I$ represents a family of curves in $C^m$ **through a point** $p \in C^m$ if, for every $l \in L$, $I(p, l)$ holds.

We often say $L$, instead of $I$, represents a family of curves, and when the context is clear, we identify every $l \in L$ with the set $\{p : I(p, l)\}$.

**Exercise 3.8.2.** *Prove that for a faithful family $L$,*

*1. for any closed $S \subseteq P$ with $\dim S \leq 1$ (curve), for a generic $l \in L$*

$$\dim I(P, l) \cap S < 1;$$

*2. for any generic point $q \in C^m$*

$$\dim I(q, L) = \dim L - 1.$$

**Remark 3.8.3.** By removing a small (proper closed) subset of points, we may always assume that every point belongs to a curve and no point belongs to almost all curves; that is, $\dim I(q, L) < \dim L$. The inverse is obvious by Zariski axioms.

For the rest of the discussion, we add the assumption that the Zariski structure on $C$ is **non-linear** (equivalently, non-locally modular) in the sense of Section B.1.3. By the results in that section, non-linearity is equivalent to the assumption that the Zariski geometry on $C$ is **ample**:

(AMP) There is a two-dimensional, irreducible, faithful family $L$ of curves on $C^2$. $L$ is locally isomorphic to an open subset of $C^2$.

**Exercise 3.8.4.** *Given $I$ and $L$ as before, show that for any generic pair of points $p_1, p_2$ from $C^2$ there are finitely many lines from $L$ through $p_1$ and $p_2$.*

Suppose we fix $\langle a, b \rangle$ generic in $C^2$.
Then

$$L_{\langle a, b \rangle} = I(L, \langle a, b \rangle)$$

represents a family of curves on $C^2$ through $\langle a, b \rangle$ (with the incidence relation $I_{\langle a, b \rangle} = I \cap (L_{\langle a, b \rangle} \times C^2)$.

By these assumptions, $\dim L_{\langle a, b \rangle} = 1$. By the smoothness theorem, Theorem 3.6.30, we can choose a one-dimensional irreducible smooth $G \subseteq L_{\langle a, b \rangle}$ which, along with the appropriate incidence relation, represents a family of curves through $\langle a, b \rangle$. Thus we have proved the following Lemma.

**Lemma 3.8.5.** *There exists an irreducible, faithful, one-dimensional smooth family N of curves through* $\langle a, b \rangle$.

**Remark 3.8.6.** Notice that once $\langle a, b \rangle$ was fixed and $N$ defined, $a, b$ become 0-definable and in particular cease to be generic.

**Definition 3.8.7.** Let $\langle a, b \rangle$ be a point in $C^2$. A subset $\gamma \subseteq \mathcal{V}_{\langle a,b \rangle}$ is said to be a **branch of a curve at** $\langle a, b \rangle$ if there are $m \geq 2$, $c \in C^{m-2}$, an irreducible smooth family $G$ of curves through $\langle a, b \rangle^\frown c$ with an incidence relation $I$ and a curve $g \in G$ such that the cover $I$ of $G \times C$,

$$\langle u, \langle x, y \rangle^\frown z \rangle \mapsto \langle u, x \rangle,$$

is regular (hence finite) and unramified at $\langle g, \langle a, b \rangle^\frown c \rangle$, and

$$\gamma = \{\langle x, y \rangle \in \mathcal{V}_{\langle a,b \rangle} : \exists z \in \mathcal{V}_c \ \langle g', \langle x, y \rangle^\frown z \rangle \in I\}$$

for a $g' \in \mathcal{V}_g \cap G(^*M)$.

The definition says that $g'$ is an infinitesimal piece of a possibly 'non-standard' curve in the neighbourhood of a nice standard $g$ passing through a standard point $\langle a, b \rangle^\frown c$.

We usually denote $\gamma$ by $\tilde{g}'$.

It follows from the definition and Proposition 3.6.2 that $\tilde{g}'$ is a graph of a function from $\mathcal{V}_a$ onto $\mathcal{V}_b$.

We call the corresponding object the **(local) function** $\tilde{g} : \mathcal{V}_a \to \mathcal{V}_b$ **(from $a$ to $b$) from a family $G$ with trajectory $c$**.

**Example 3.8.8.** Let $C = \mathbb{C}$ be the field of complex numbers. Consider the standard map from Example 2.2.4 as a partial specialisation from $^*\mathbb{C}$ onto $\mathbb{C}$.

Let $L$ be the family of curves in $\mathbb{C}^3$ given by

$$I = \{ux + v(y - 1) + z(z - 1) = 0 \ \& \ ux^2 + v(y - 1)z + z(z - 1) = 0\} \subseteq \mathbb{C}^5.$$

For each choice of $u, v \in \mathbb{C}$, $u \neq 0$ or $v \neq 0$, the curve $g_{u,v} = I(u, v, \mathbb{C}^3)$ passes through the points $(x, y, z) = (0, 1, 0)$ and $(x, y, z) = (0, 1, 1)$.

The projections of the $g_{u,v}$'s on the $(x, y)$-plane are curves through $(0, 1)$ given by the equation

$$uv^2x(y-1)^2 - v^3(y-1)^3 + u^2x^2(x-1)^2 + uv(x-1)x(y-1) = 0 \quad (3.4)$$

with a nodal singularity in $(0, 1)$.

On the other hand $g_{u,v}$ is non-singular in both $(0, 1, 0)$ and $(0, 1, 1)$ and so defines two families of local functions $\tilde{g}^0_{u,v} : \mathcal{V}_0 \to \mathcal{V}_{1,0}$ and $\tilde{g}^1_{u,v} \mathcal{V}_0 \to \mathcal{V}_{1,1}$.

The first coordinate of the functions defines the branches of the planar curves (3.4) through $(0, 1)$, with the corresponding trajectory $z = 0$ and $z = 1$.

**Lemma 3.8.9.** *Given an irreducible, faithful, smooth family $G$ of curves through $\langle a, b \rangle$ and $g_1, g_2 \in {}^*G$, if $\tilde{g}_1 = \tilde{g}_2$ as functions $V_a \to V_b$, then $g_1 = g_2$. In other words, $G$ is represented faithfully by local functions.*

*Proof.* This is immediate by assumption (iii) on the family of curves. □

**Definition 3.8.10.** Let $I_1$ and $I_2$ be two families of local functions from $a$ to $b$, with trajectories $c_1$ and $c_2$. We say that the corresponding **branches defined by $g_1 \in G_1$ and $g_2 \in G_2$ are tangent at $\langle a, b \rangle$**, and write

$$g_1 \, T \, g_2$$

if there is an irreducible component $S = S_{(I_1, I_2, a, b, c_1, c_2)}$ of the set

$$\{\langle u_1, u_2, x, y, z_1, z_2 \rangle \in G_1 \times G_2 \times C^2 \times C^{m_1 - 2} \times C^{m_2 - 2} : \\ \langle u_1, x, y, z_1 \rangle \in I_1 \ \& \ \langle u_2, x, y, z_2 \rangle \in I_2 \} \tag{3.5}$$

such that

1. $\langle g_1, g_2, a, b, c_1, c_2 \rangle \in S$;
2. the image of the natural projections of $S$ into $G_1 \times G_2$

$$\langle u_1, u_2, x, y, z_1, z_2 \rangle \mapsto \langle u_1, u_2 \rangle$$

   is dense in $G_1 \times G_2$; and
3. for $i = 1$ and $i = 2$ the images of the maps

$$\langle u_1, u_2, x, y, z_1, z_2 \rangle \mapsto \langle x, y, z_i, u_i \rangle$$

are dense in $I_i$ and the corresponding coverings by $S$ are regular at the points $\langle a, b, c_i, g_i \rangle$.

**Remark 3.8.11.** Once $I_1, I_2, a, b, c_1, c_2$ have been fixed, one has finitely many choices for the irreducible component $S$.

**Remark 3.8.12.** We can write item 3 as a first-order formula

$$\langle a, b, c_i, g_i \rangle \in \text{reg}\,(S/I_i).$$

**Corollary 3.8.13.** *The formula*

$$T := \bigcup_S S(u_1, u_2, a, b, c_1, c_2) \ \& \ \langle a, b, c_1, u_1 \rangle \in \text{reg}\,(S/I_1)$$

$$\& \langle a, b, c_2, u_2 \rangle \in \text{reg}\,(S/I_2)$$

*(with parameters $a, b, c_1, c_2$) defines the tangency relation between $u_1 \in G_1$ and $u_2 \in G_2$.*

**Proposition 3.8.14.** *Given $G_1, G_2$ families of curves defining local functions from a to b, $g_1 \in G_1$, generic in $G_1$, and $g_2 \in G_2$ generic in $G_2$, the following three conditions are equivalent:*

1. $g_1 \, T \, g_2$;
2. $\forall x \in V_a \; \forall g_1' \in V_{g_1} \; \exists g_2' \in V_{g_2} : \tilde{g}_1'(x) = \tilde{g}_2'(x)$; *and*
3. $\forall x \in V_a \; \forall g_2' \in V_{g_2} \; \exists g_1' \in V_{g_1} : \tilde{g}_1'(x) = \tilde{g}_2'(x)$.

*Moreover, there are Zariski-open subsets $G_1^0 \subseteq G_1$ and $G_2^0 \subseteq G_2$ such that the three conditions are equivalent for any $g_1 \in G_1^0$ and $g_2 \in G_2^0$.*

*Proof.* Suppose Proposition 3.8.14 (1) holds. Let $S = S_{(I_1, I_2, a, b, c_1, c_2)}$ define the tangency of $g_1, g_2$.

Because $S$ is a covering of $I_1$ regular at $\langle a, b, c_1, g_1 \rangle$ by Definition 3.8.10 (3), Proposition 3.6.2 gives us

$$\forall \langle x, y, z_1, g_1' \rangle \in V_{a,b,c_1,g_1} \cap I_1 \;\; \exists \langle z_2, g_2' \rangle \in V_{\langle c_2, g_2 \rangle} \cap C^{m_2 - 2} \times G_2 :$$
$$\langle g_1', g_2', x, y, z_1, z_2 \rangle \in S.$$

This immediately implies Proposition 3.8.14 (2).

Proposition 3.8.14 (3) follows from Proposition 3.8.14 (1) and similarly when we consider $I_2$ instead of $I_1$.

Conversely, assume Proposition 3.8.14 (2). Choose $\langle x, y, z_1, g_1' \rangle$ generic in $V_{\langle a, b, c_1, g_1 \rangle} \cap I_1$. By Proposition 3.8.14 (2), $\tilde{g}_1'(x) = y = \tilde{g}_2'(x)$ for some $g_2' \in V_{g_2} \cap G_1$ as a branch with the trajectory $c_2$; that is, $\langle x, y, z_2, g_2' \rangle \in I_2$ for some $z_2 \in V_{c_2}$. Notice that $\langle x, y \rangle$ is generic in $C^2$ by our choice. By Lemma 3.6.3, with $I_2$ in place of $F$ and an open subset of $C^2$ in place of $D$, we can choose $\langle z_2, g_2' \rangle \in V_{\langle c_2, g_2 \rangle} \cap C^{m_2 - 2} \times G_2$ of maximal possible dimension, that is with $\mathrm{cdim}\,(g_2'/x, y, z_1, M) = \dim G_2 - 1$. We thus have

$$\mathrm{cdim}\,(\langle g_1', g_2' \rangle) = \mathrm{cdim}\,(\langle g_1', g_2', x, y, z_1, z_2 \rangle) = \dim G_1 + \dim G_2. \quad (3.6)$$

Let $S$ be the locus of $\langle g_1', g_2', x, y, z_1, z_2 \rangle$ over $M$. By the dimension calculations in Equation (3.6), we see that $S$ is an irreducible component of the set 3.8.10 (3.5).

By applying the specialisation to the point $\langle g_1', g_2', x, y, z_1, z_2 \rangle$, we see that $\langle g_1, g_2, a, b, c_1, c_2 \rangle \in S$. Also, $S$ projects on dense subsets of $I_1$ and $I_2$ by construction. By Equation (3.6), $S$ projects on a dense subset of $G_1 \times G_2$. Finally, Definition 3.8.10 (3) follows from the following:

Claim: For $i = 1, 2$, the point $\langle a, b, c_i, g_i \rangle$ belongs to $\mathrm{reg}(S/I_i)$.

Indeed, because $g_i$ is generic in $G_i$,

$$\operatorname{cdim}(\langle a, b, c_i, g_i \rangle / a, b, c_i) = \dim G_i$$

and, by Lemma 3.5.13,

$$\dim(I_i \setminus \operatorname{reg}(S/I_i)) \leq \dim I_i - 2 = \dim G_i - 1.$$

So, $S$ defines tangency of $g_1$ and $g_2$ and Proposition 3.8.14 (1) follows. Symmetrically we get Proposition 3.8.14 (1) from 3.8.14 (3).

It remains to prove the 'moreover' clause of the proposition. This is immediate after one notices that the only time we use the assumption of $g_1$ and $g_2$ as generic is in the proof of this claim. Here the assumption can be replaced by choosing $G_i^0$ so that $\{a\} \times G_i^0 \subseteq \operatorname{reg}(S/I_i)$ for each of the finitely many possible $S$. $\qquad\square$

**Corollary 3.8.15.** *Tangency is a reflexive binary relation on an open subset of curves through $\langle a, b \rangle$; that is, $gTg$ holds for any $g \in G^0 \subseteq G$ of a faithful, smooth family $G$.*

**Proposition 3.8.16.** *Let $G_i$ be families of branches of curves through $\langle a, b \rangle$ with trajectories $c_i$, and let $g_i \in G_i$ generic, $i = 1, 2, 3$. Assume $g_1 T g_2$ and $g_2 T g_3$ hold. Then $g_1 T g_3$.*

*Moreover, for some open subsets $G_i^0 \subseteq G_i$, $i = 1, 2, 3$,*

$$\forall g_1, g_2, g_3 : g_1 \in G_1^0 \,\&\, g_2 \in G_2^0 \,\&\, g_3 \in G_3^0 \,\&\, g_1 T g_2 \,\&\, g_2 T g_3 \Rightarrow g_1 T g_3.$$

*Proof.* We may use Proposition 3.8.14 (2) as the definition of tangency, which immediately implies the required. $\qquad\square$

**Proposition 3.8.17.** *Let $G_1, G_2$ be smooth, faithful families of branches of curves through $\langle a, b \rangle$, and let $g_1$ be generic in $G_1$. Then the set $g_1 T$ of curves in $G_2$ tangent to $g_1$ is at most of dimension $\dim G_2 - 1$.*

*If the tangency relation $T \subseteq G_1 \times G_2$ projects densely on $G_1$ or $G_2$, then $\dim T = \dim G_1 + \dim G_2 - 1$. In particular, this is the case when $G_1 = G_2$.*

*Proof.* Because $I_1 g_1$ is a curve in $C^{m_1}$ passing through $\langle a, b, c_1 \rangle$, there is $\langle a', b', c_1' \rangle \in V_{\langle a, b, c_1 \rangle} \cap^* C^{m_2}$ generic over $g_1$. Let $D$ be the curve in $C^2$ defined by $g_1$, that is, the closure of the projection of $I_1 g_1$ on $C^2$. By construction $\langle a, b \rangle, \langle a', b' \rangle \in D$. We can consider an irreducible component of $D$ containing $\langle a', b' \rangle$ (and so $\langle a, b \rangle$) rather than $D$, so we assume $D$ is irreducible.

The projection $\text{pr}_{C^2} I_2$ of the irreducible set $I_2$ contains $\langle a, b \rangle$ and is dense in $C^2$. Hence the closed set

$$F = \{\langle x, y, z_2, u_2 \rangle \in I_2 : \langle x, y \rangle \in D\}$$

is a cover (possibly reducible) of an open subset of $D$ containing $\langle a', b' \rangle$. Obviously,

$$\dim F = \dim I_2 - 1 = \dim G_2,$$

and by the addition formula the dimension of generic fibre is

$$\dim F_{\langle a', b' \rangle} = \dim I_2 - 2 = \dim G_2 - 1.$$

Let

$$F = F^0 \cup \cdots \cup F^k$$

be the decomposition of $F$ into irreducible components. Because $\{\langle a, b, c_2 \rangle\} \times G_2 \subseteq F$ and is of the same dimension as $F$, we have that $\{\langle a, b, c_2 \rangle\} \times G_2$ is a component of $F$, say $F^0$. This is the only component which projects into a point, namely $\langle a, b \rangle$.

Any other component $F^i$ is a covering of an open subset of $D$ and so by Corollary 3.5.14 is regular at every point. In particular, its fibre over $\langle a, b \rangle$, $F^i(\langle a, b \rangle, y)$ is either empty or of generic dimension, that is, equal to $\dim G_2 - 1$. Let

$$E_{g_1}(a, b, y) = F^1(a, b, y) \vee \cdots \vee F^k(a, b, y).$$

*Claim 1.* For any $h \in G_2$ such that $\neg E_{g_1}(a, b, \langle c_2, h \rangle)$, there is no $h' \in V_h \cap G_2$ and $c_2' \in V_{c_2}$ with $\langle a', b', c_2', h' \rangle \in I_2$.

Indeed, if these exist, then $\langle a', b', c_2', h' \rangle \in F$ and moreover $\langle a', b', c_2', h' \rangle \in F^i$ for some $i = 1, \ldots, k$. By applying the specialisation $\pi$ and remembering that $F^i$ is closed in $F$, we get $\langle a, b, c_2, h \rangle \in F^i$. This contradicts the choice of $h$ and proves the claim.

In particular, the type

$$\text{Nbd}_{c_2, h}(y) \cup \{F(a', b', y)\}$$

(see Definition 2.2.19) is inconsistent by Lemma 2.2.21. This means that

$$\models F(a', b', y) \rightarrow Q(e', y) \tag{3.7}$$

for some $e$ in $M$, $e' \in V_e$ and Zariski-closed $Q$ such that $\neg Q(e, \langle c_2, h \rangle)$ holds. In particular, if $Q(e, y)$ is consistent,

$$\dim G_2 - 1 = \dim F(a', b', y) \leq \dim Q(e, y) < \dim G_2,$$

so

$$\dim Q(e, y) = \dim G_2 - 1. \tag{3.8}$$

By the assumption of the proposition, there is $g_2 \in G_2$ tangent to $g_1$. By Corollary 3.8.13, we have that $F(a', b', c_2', g_2')$, for some $g_2' \in V_{g_2}$ and $c_2' \in V_{c_2}$, and $E_{g_1}(a, b, \langle c_2, g_2 \rangle)$ hold. So, $E_{g_1}(a, b, y)$ can be taken as an example of a consistent $Q(e, y)$ ($e = \langle a, b \rangle$) for which Equation (3.7) holds.

Choose now Zariski-closed $Q(e, y)$ ($e$ in $M$) minimal among those satisfying condition (3.7) and $Q(e, \langle c_2, g_2 \rangle)$. So, Equation (3.8) holds for such $Q(e, y)$.

*Claim 2.* For any $h$ such that $Q(e, \langle c_2, h \rangle)$ holds, the type

$$\mathrm{Nbd}_{c_2, h}(y) \cup \{F_{\langle a', b' \rangle}(y)\}$$

is consistent.

Indeed, otherwise for some $f$ in $M$, $f' \in V_f$ and closed $R(f, y)$ such that $\neg R(f, \langle c_2, h \rangle)$ holds, we have $\models F(a', b', y) \rightarrow R(f', y)$. In particular, $R(f', \langle c_2', g_2' \rangle)$ and $R(f, \langle c_2, g_2 \rangle)$ hold. Hence, $\dim Q(e', y) \,\&\, R(f', y) \geq \dim G_2 - 1$. Hence,

$$\dim Q(e, y) \,\&\, R(f, y) = \dim G_2 - 1.$$

By minimality $Q(e, y) \,\&\, R(f, y) \equiv Q(e, y)$ and so $M \models R(f, \langle c_2, g_2 \rangle)$, which is a contradiction.

The claim implies, by Lemma 2.2.21, that there is $\langle c_2', g_2' \rangle \in V_{\langle c_2, g_2 \rangle}$ generic in $F_{\langle a', b' \rangle}(y)$ and so $\langle a', b', c_2', g_2' \rangle \in I_2$.

Let $S$ be the locus of $\langle a', b', c_1', c_2', g_1, g_2' \rangle$ over $\emptyset$. Taking into account that $F_{\langle a', b' \rangle}(y)$ is defined over $\{a, b, c_1, c_2, g_1 a', b'\}$ and the fact that $g_1, a', b'$ have been chosen generically over $ab$, we get by dimension calculations

$$\mathrm{cdim}\, \langle a', b', c_1', g_1 \rangle = \dim I_1$$

$$\mathrm{cdim}\, \langle a', b', c_2', g_2' \rangle = \dim I_2$$

$$\dim G_1 + \dim G_2 = \mathrm{cdim}\, \langle a', b', c_2', c_2', g_1, g_2' \rangle = \mathrm{cdim}\, \langle g_1, g_2' \rangle,$$

which show that $S$ satisfies Definition 3.8.10 (2) and Definition 3.8.10 (3). By construction, Definition 3.8.10 (1) holds as well. So, $g_1$ is tangent to $g_2$ via $S$.

Obviously, we can choose $g_1 \in G_1$ generic iff $T$ projects generically on $G_1$. Claim 2 allows us to choose $g_2$ so that $\mathrm{cdim}\,(g_2/g_1) \geq \dim G_2 - 1$; hence, $\mathrm{cdim}\,(g_1, g_2) \geq \dim G_1 + \dim G_2 - 1$. However, claim 1 does not allow $g_2$ to be generic over $g_1$, so

$$\mathrm{cdim}\,(g_1, g_2) = \dim G_1 + \dim G_2 - 1.$$

Finally, because tangency is reflexive on an open subset of $G$, we have the dimension equality in the case $G_1 = G_2 = G$. □

We can now draw the following picture. Given a point $\langle a, b \rangle \in C^2$ we have, for some trajectories $c \in C^m$, $m \in \mathbb{N}$, definable irreducible smooth families $G_{c,i}$ (sorts) of branches (of curves) through $\langle a, b \rangle$. The tangency relation $T$ between elements of

$$\mathcal{G}_{\langle a,b \rangle} = \bigcup G_{c,i}$$

is definable when restricted to any pair of sorts, $T \cap (G_{c_1,i_1} \times G_{c_2,i_2})$.

The tangency properties can be summarised in the following corollary.

**Corollary 3.8.18.** *$T$ is an equivalence relation on $\mathcal{G}_{\langle a,b \rangle}$. The restriction of $T$ to any pair of sorts is closed and proper (non-trivial).*

# 4

# Classification results

Throughout this chapter, we assume unless stated otherwise that our Zariski geometry is $C$, the one-dimensional, irreducible, pre-smooth Zariski structure satisfying (EU) which we studied in Section 3.8. We follow the notation and assumptions of that sub-section. Our main goal is to classify such structures, which is essentially achieved in Theorem 4.4.1. In fact, the proof of the main theorem deepens the analogy between our abstract Zariski geometry and algebraic geometry. In particular, we prove generalisations of Chaos theorem on analytic subsets of projective varieties and of Bezout's theorem. As a by-product, we develop the theory of groups and fields living in pre-smooth Zariski structures.

## 4.1 Getting a group

Our aim in this section is to obtain a Zariski group structure living in $C$. The main steps of this construction are as follows:

- We consider the composition of local functions $\mathcal{V}_a \to \mathcal{V}_a$ (branches of curves through $\langle a, a \rangle$) modulo the tangency and show that generically it defines an associative operation on a pre-smooth Zariski set, a *pre-group of jets*.
- We prove that any Zariski pre-group can be extended to a group with a pre-smooth Zariski structure on it. This is an analogue of Weil's theorem on group chunks in algebraic geometry.

We consider tangency of branches of curves through $\langle b, a \rangle$, $\langle a, a \rangle$, and potentially through other points on $C^2$. We keep the notation $T$ for this tangency as well, when there is no ambiguity about the point at which the branches are considered. In case there is a need to specify the point, we do it by adding a subscript: $T_{\langle a,b \rangle}$, $T_{\langle a,a \rangle}$, and so on.

### 4.1.1 Composing branches of curves

**Definition 4.1.1.** Given $g \in G_c \subseteq \mathcal{G}_{\langle a,b \rangle}$, which we identify with the local function $\tilde{g} : \mathcal{V}_a \to \mathcal{V}_b$, we define the inverse

$$g^{-1} = \{(y, x) \in \mathcal{V}_b \times \mathcal{V}_a : (x, y) \in g\}$$

and define $G_c^{-1}$ to be a copy of the definable set $G_c$ with the inverse action $\mathcal{V}_b \to \mathcal{V}_a$ induced by its elements.

Obviously, $g^{-1}$ determines a branch through $\langle b, a \rangle$ with a trajectory we consider to be in inverse order to (the sequence) $c$ and write as $c^{-1}$. We denote the new family of branches $\mathcal{G}_{\langle a,b \rangle}^{-1}$.

**Lemma 4.1.2.** *For every* $g_1, g_2 \in \mathcal{G}_{\langle a,b \rangle}$,

$$g_1 T g_2 \qquad \textit{iff} \qquad g_1^{-1} T g_2^{-1}.$$

*Proof.* Use the criterion given in Proposition 3.8.14 and the fact that the local functions are bijections. $\qquad\square$

**Definition 4.1.3.** For any $g_1 \in G_{c_1}$ and $g_2 \in G_{c_1}$ with $G_{c_1}, G_{c_2} \subseteq \mathcal{G}_{\langle a,b \rangle}$, define the **composition** curve

$$g_2^{-1} \circ g_1 = \{(x_1, x_2) \in C^2 : \exists y, z_1, z_2 \ (x_1, y, z_1) \in g_1 \ \& \ (z_2, y, x_2) \in g_2^{-1}\}$$

and its branch at $\langle a, a \rangle$

$$(g_2^{-1} \circ g_1)_{\langle a,a \rangle} = \{(x_1, x_2) \in \mathcal{V}_a \times \mathcal{V}_a : x_2 = \tilde{g}_2^{-1}(\tilde{g}_1(x_1))\}.$$

By the definition of branches, if $c_1, c_2$ are the corresponding trajectories of $g_1, g_2$, then the composition $(g_2^{-1} \circ g_1)_{\langle a,a \rangle}$ is a branch of $g_2^{-1} \circ g_1$ at $\langle a, a \rangle$ with the trajectory $c_1 b c_2^{-1}$. We denote the new family of branches $\mathcal{G}_{\langle a,b \rangle}^{-1} \circ \mathcal{G}_{\langle a,b \rangle}$.

More generally, given two families of branches $\mathcal{G}_{\langle a,b \rangle}$ and $\mathcal{G}_{\langle b,d \rangle}$ through $\langle a, b \rangle$ and $\langle b, d \rangle$, respectively, we consider the family $\mathcal{G}_{\langle b,d \rangle} \circ \mathcal{G}_{\langle a,b \rangle}$ of branches of curves defined as compositions of the corresponding local functions.

**Lemma 4.1.4.** *$T$ is preserved by composition of branches of sufficiently generic pairs of curves. Namely, there is an open subset $V \subseteq \mathcal{G}_{\langle a,b \rangle} \times \mathcal{G}_{\langle a,b \rangle}$ such that, for any $\langle g_1, g_2 \rangle, \langle h_1, h_2 \rangle \in V$,*

$$g_1 T h_1 \ \& \ g_2 T h_2 \Rightarrow g_2^{-1} \circ g_1 \ T \ h_2^{-1} \circ h_1.$$

*Proof.* The proof is the same as that of Lemma 4.1.2. $\qquad\square$

Notice that by definition $g^{-1}$ and $g_2^{-1} \circ g_1$ are branches of curves represented by members of families of curves $G$ and $G_1 \times G_2$, respectively, through correspondent fixed trajectories ($c^{-1}$ for $g^{-1}$ and $c_1 b c_2^{-1}$ for $g_2^{-1} \circ g_1$). It is also important to notice the following Lemma.

**Lemma 4.1.5.** *Given smooth and faithful families of branches of curves* $G_1$ *and* $G_2$, *the families* $G_1^{-1}$ *and* $G_1^{-1} \circ G_2$ *are smooth and faithful.*

*Proof.* The statement for $G_1^{-1}$ is obvious. Consider the second kind of family.

By definition, the family $G_1^{-1} \circ G_2$ is represented by an incidence relation between points of the smooth set $G_1 \times G_2$ and $C^2 \times C^m$, some $m$ determined by the trajectories. Hence, it is smooth.

For faithfulness, we need to check that for generic $\langle g_1, h_1, g_2, h_2 \in G_1^2 \times G_2^2 \rangle$, the intersection $g_2^{-1} \circ g_1 \cap h_2^{-1} \circ h_1$ is finite. Suppose it is not.

Notice first that $g_2^{-1} \circ g_1$ is a one-dimensional subset of $C^2 \times C^m$. By applying $h_2$, we get $h_2 \circ g_2^{-1} \circ g_1$, a one-dimensional subset definable over $\langle h_2, g_2, g_1 \rangle$ and intersecting the curve $h_1$ in a one-dimensional irreducible component. This holds for every generic over $\langle h_2, g_2, g_1 \rangle$ element of $G_2$. It follows that a generic pair $h_2, h_2'$ from $G_2$ has an infinite intersection, a contradiction with faithfulness of $G_2$. $\qquad\square$

The lemma shows that the properties of tangency of Section 3.8 are applicable, in particular that $T$ is a definable relation on $\mathcal{G}_{\langle a, a \rangle}$.

Now we recall the smooth one-dimensional family of curves $N$ through $\langle a, b \rangle$ introduced in Lemma 3.8.5. By our definitions and the facts established previously, $N^{-1} \circ N$ can be considered a smooth faithful family of branches of curves through $\langle a, a \rangle$ with trajectory $b$. The family of branches will be denoted $H_{aa}$. We also consider the family of curves denoted $(N^{-1} \circ N) \circ (N^{-1} \circ N)$ and defined as compositions of pairs of curves of $N^{-1} \circ N$. Correspondingly, this parametrises the family of branches through $\langle a, a \rangle$ with trajectory $\langle a, b, a \rangle$ which we denote $H_{aa} \circ H_{aa}$. To keep the nice behaviour of $T$, we restrict ourselves to an open subset $(H_{aa} \circ H_{aa})^0$ of $H_{aa} \circ H_{aa}$ for which the equivalences of Proposition 3.8.14 hold.

**Lemma 4.1.6.** *Let* $(n, \ell_1, \ell_2) \in N^3$ *be generic. Then*

*(1)* $n^{-1} \circ \ell_1$ *is not tangent to* $n^{-1} \circ \ell_2$ *and*
*(2)* $\ell_1^{-1} \circ n$ *is not tangent to* $\ell_2^{-1} \circ n$.

*Proof.* By **Lemma 4.1.2**, the two statements are equivalent to each other. Suppose both (1) and (2) are negated. Then, because the tangency of generic **germs** is witnessed by a closed relation, this holds for any generic triple. Take any generic string of elements of $N$: $n_1, m_1, m_2, n_2$. By our assumptions,

$m_1^{-1} \circ n_1 \, T \, m_1^{-1} \circ n_2$, $m_1^{-1} \circ n_2 \, T \, m_2^{-1} n_2$, so by transitivity $m_1^{-1} \circ n_1$ is tangent to $m_2^{-1} \circ n_2$, contradicting the genericity, by Proposition 3.8.17. $\qquad \square$

**Lemma 4.1.7.** *Given a generic triple* $\langle \ell_1, \ell_2, n_1 \rangle \in N^3$, *there is* $n_2 \in N$ *such that* $n_1^{-1} \circ \ell_1 \, T \, n_2^{-1} \circ \ell_2$.

*Moreover,* $n_2 \in \mathrm{acl}(\ell_1, \ell_2, n_1)$ *and every three of the four elements* $\ell_1, \ell_2, n_1, n_2$ *are independent.*

*Proof.* By Proposition 3.8.17, the tangency class of $n_1^{-1} \circ \ell_1$ in $H_{aa}$ is a one-dimensional set in $N^2$. Choose $n^{-1} \circ \ell$ tangent to $n_1^{-1} \circ \ell_1$ and such that $\mathrm{cdim}\,(\langle n, \ell \rangle / \{n, \ell_1\}) = 1$. Suppose, towards a contradiction, that $\mathrm{cdim}\,(\ell / \{n_1, \ell_1\}) = 0$; that is, $\ell \in \mathrm{acl}(n_1, \ell_1)$. Then this is true of every pair $\langle n, \ell \rangle$ of the same type over $n_1, \ell_1$. Then there is an $\ell \in N$ such that $n_2^{-1} \circ \ell$ is tangent to $n_1^{-1} \circ \ell_1$ for almost all $n_2 \in N$. Hence, without loss of generality we may assume that $n = n_1$. This clearly contradicts Lemma 4.1.6.

Thus, we have proven that $\ell$ is generic in $N$ over $\{n_1, \ell_1\}$. Because any two generic elements are of the same type, we may choose $\ell = \ell_2$.

Symmetrically, $n = n_2$ is generic in $N$ over $\{n_1, \ell_1\}$. Notice also that $n_2 \in \mathrm{acl}(\ell_1, \ell_2, n_1)$ and $\ell_2 \in \mathrm{acl}(\ell_1, n_1, n_2)$ by dimension data. This proves the independence statement. $\qquad \square$

**Lemma 4.1.8.** *Given a generic* $\langle f_1, f_2 \rangle \in H_{aa} \times H_{aa}$, *there is generic* $g \in H_{aa}$ *such that* $g$ *is tangent to the composition* $f_1 \circ f_2$.

*Proof.* Fix an $\ell \in N$ generic over $\langle f_1, f_2 \rangle$. By Lemma 4.1.7, there are $n_1, n_2 \in N$ such that

$$f_1 \, T \, n_1^{-1} \circ \ell \text{ and } f_2 \, T \, \ell^{-1} \circ n_2. \tag{4.1}$$

*Claim.*

$$f_1 \circ f_2 \, T \, n_1^{-1} \circ n_2.$$

To prove the claim we use Proposition 3.8.14. Notice that $\langle n_1, n_2 \rangle$ is a generic pair, by the last part of Lemma 4.1.7.

Consider $x \in V_a$ and $\langle n_1', n_2' \rangle \in V_{(n_1, n_2)}$. We need to find $\langle f_1', f_2' \rangle \in V_{(f_1, f_2)}$ such that

$$\tilde{f}'_1 \circ \tilde{f}'_2(x) = \tilde{n}'_1{}^{-1} \circ \tilde{n}'_2(x).$$

By Equation (4.1), we choose $f_2' \in V_{f_2}$ so that $f_2'(x) = y = \ell^{-1} \circ n_2'(x)$. Next we choose $f_1' \in V_{f_1}$ so that $f_1'(y) = n_1'^{-1} \circ \ell(y)$. These satisfy our requirement and prove the claim.

## 4.1.2 Pre-group of jets

We are now ready to prove the main idea for establishing the existence of a group structure called the **group of jets on** $C$. The next proposition and theorem are Z-analogues of the theory presented in the purely model-theoretic context in Chapter 5 of Poizat (1987).

**Proposition 4.1.9 (pre-group of jets).** *There is a one-dimensional irreducible manifold $U$ and a constructible irreducible ternary relation $P \subseteq U^3$ which is the graph of a partial map $U^2 \to U$ and determines a **partial Z-group structure** on $U$; that is, there is an open subset $V \subseteq U^2$ such that*

(i) *for any pair* $\langle u, v \rangle \in V$, *there is a unique* $w = u * v \in U$ *satisfying* $\langle u, v, w \rangle \in P$;

(ii) *for any generic* $\langle u, v, w \rangle \in U^3$,

$$u * (v * w) = (u * v) * w;$$

(iii) *for each pair* $\langle u, v \rangle \in V$, *the equations*

$$u * x = v \text{ and } y * u = v$$

*have solutions in $U$.*

*Proof.* We start with the $N$ of Lemma 3.8.5 and consider the smooth set $N \times N$ $(N^{-1} \circ N)$ as a family of curves $H_{aa}$. The equivalence relation $T$ by dimension calculations of 3.8.17 has classes of dimension 1 on an open subset $H$ of $N^2$ and $\dim H/T = 1$. Because $N$ is locally isomorphic to $C$ and $C$ is ample, there must be an irreducible curve $S \subseteq H$ which intersects with infinitely many classes of equivalence $T$, so each intersection is finite. At the cost of deleting finitely many points, we can assume that $S$ is smooth. Hence, by Proposition 3.7.22, $U = S/T$ is a manifold. Because $\dim H/T = 1 = \dim S/T$ and the set on the right is irreducible, we can choose $H$ so that

$$S/T = H/T = U.$$

The composition of branches preserves tangency and hence the partial map of

$$S/T \times S/T \to (S \times S)^0/T \subseteq U \times U$$

is well defined generically, with some open $(S \times S)^0 \subseteq S \times S$. This map can be equivalently interpreted as

$$H_{aa}/T \times H_{aa}/T \to (H_{aa} \circ H_{aa})^0/T,$$

with some open $(H_{aa} \circ H_{aa})^0 \subseteq H_{aa} \circ H_{aa}$.

Lemma 4.1.8 identifies $(H_{aa} \circ H_{aa})^0$ with an open subset of $H_{aa}/T$. This gives us a continuous map

$$* : U^2 \to U$$

defined on an open subset $V$ of $U^2$ and with the image of a one-dimensional subset of $U$. This proves part (i).

Part (ii) follows from the fact that the operation $*$ corresponds to the composition of local functions.

(iii) Because generic pair $\langle u, x \rangle$ is sent by $*$ to a generic element of $U$ (see Lemma 4.1.8 again), for every generic pair $\langle u, v \rangle$, the equation

$$u * x = v \tag{4.2}$$

has a solution in $U$. So, it holds for an open subset of $U^2$. By symmetry, the same is true for $y * u = v$. □

**Definition 4.1.10.** Let $G$ be a manifold and $P \subseteq G^3$ a closed ternary relation which is the graph of a binary operation

$$(u, v) \mapsto u \cdot v$$

on $G$. We say that $G$ is a **Z-group** if $(G, \cdot)$ is a group.

A group $G$ with dimension is called **connected** if the underlying set $G$ cannot be represented as

$$G = S_1 \cup S_2, \quad S_1 \cap S_2 = \emptyset, \quad \dim S_1 = \dim S_2.$$

This definition is applicable to groups with superstable theories and is known to be equivalent to the condition that $G$ *has no proper definable subgroups of finite index*. Of course, in the Zariski context 'definable' can be replaced by 'constructible'.

If, for a Z-group $G$, the underlying set $G$ is irreducible, then $G$ is obviously connected. Conversely, Lemma 4.1.11 applies.

**Lemma 4.1.11.** *A connected Z-group $G$ is irreducible.*

*Proof.* It follows from the definition of connectedness that $G$ contains a dense open subset, say $U$. For any $g \in G$, we can find an $h \in G$ so that $g \in h \cdot U$, and obviously $h \cdot U$ is a dense open subset. So, we can cover $G$ by a union of dense open subsets of the form $h \cdot U$. By Noetherianity, there is a finite subcover of this cover. It follows by Proposition 3.7.16 that $G$ is irreducible. □

**Exercise 4.1.12.** *Suppose G is connected. Prove the following:*

   (i) *If $H \subseteq G$ is a finite subset with the property $g^{-1}Hg = H$, then $H \subseteq C(G)$, the centre of G.*

  (ii) *$G/C(G)$ is connected and has no finite normal subsets.*

 (iii) *If $C(G)$ is finite, then $G/C(G)$ is not Abelian.*

 (iv) *If G is connected and $\dim G = 1$, then for any $a_1, a_2 \in G \setminus C(G)$ there is $g \in G$ such that $g^{-1}a_1g = a_2$.*

**Theorem 4.1.13 (Z-version of Weil's theorem on pre-groups).** *For any partial irreducible Z-group U, there is a connected Z-group G and a Z-isomorphism between some dense open $U' \subseteq U$ and dense open $G' \subseteq G$.*

*Proof.* The proof is similar to that in Bouscaren's work (1989). We use the notation of Proposition 4.1.9 without assuming that $U$ is one-dimensional.

Because projections are open maps, we may assume that the projections of $V$ on both coordinates are equal to $U$.

We define $G$ to be a semi-group of partial functions $U \to U$ generated by shifts by elements $a \in U$:

$$s_a : u \mapsto a * u.$$

We consider two elements $h, g \in G$ equal if $h(u) = g(u)$ on an open subset of $U$. The semi-group operation is defined by the composition of functions.

*Claim 1.* For every $a, b, c \in U$, there are $d, f \in U$ such that

$$a * (b * (c * u))) = d * (f * u))$$ on an open subset of $U$.

*Proof.* By Proposition 4.1.9(iii), we can find $b', b'' \in U$ such that $b' * b'' = b$, and each of them is generic in $U$ over $a, b, c$. Then

$$a * (b * (c * u))) = a * ((b' * b'') * (c * u))) = a * (b' * (b'' * (c * u))),$$

because $b'' * (c * u)$ and $b' * (b'' * (c * u))$ are well defined. Because $a, b'$ and $b'' * (c * u)$ are independent generics of $U$, we can continue

$$a * (b' * (b'' * (c * u))) = (a * b') * (b'' * (c * u)).$$

Also $b''$ and $c$ are independent and $b''$ is generic, so $b'' * c = f \in U$ is defined. Because $u$ is generic over the rest, we have $b'' * (c * u) = f * u$. Letting $a * b' = d$, we finally have $a * (b * (c * u))) = d * (f * u))$ for $u$ generic over $a, b, c, d, f$, which proves the claim.

As a corollary of the claim, we can identify $G$ with the constructible sort $(U \times U)/E$ where $E$ is an equivalence relation

$$\langle a_1, a_2 \rangle \, E \, \langle b_1, b_2 \rangle \quad \text{iff} \quad a_1 * (a_2 * u) = b_1 * (b_2 * u) \text{ on an open subset of } U.$$

The last condition can be replaced by the condition

$$\dim\{u : \exists v, w, x \in U \ P(a_2, u, v) \, \& \, P(a_1, v, x) \, \& \, P(b_2, u, w) \, \& \, P(b_1, w, x)\}$$
$$= \dim U$$

and so is constructible.

We also should notice that there is a natural embedding of shifts $s_a$, $a \in U$, into $G$; just consider $a = a' * a''$ for $a', a'' \in U$.

The latter also gives us an embedding of the pre-group $U$ into the semi-group $G$.

*Claim 2.* $G$ is a group.

In fact, this is a general fact about semi-groups without zero being definable in stable structures. If $g \in G$ does not have an inverse, then

$$g^{n+1} G \subsetneq g^n G, \qquad \text{for all } n \in \mathbb{Z}_{>0}.$$

This defines the strict order property on $M$, contradicting stability.

We have now the group $G$ generated by its subset $U$ as $U \cdot U = G$.

*Claim 3.* Let $d = \dim U^2$. There is a finite subset $\{a_1, \ldots, a_d\}$ of $U$ such that

$$G = \{a_1, \ldots, a_d\} \cdot U^{-1} = \bigcup_{i=1}^{d} \{a_i \cdot v : v \in U\}.$$

*Proof.* Choose $a_1, \ldots, a_d \in U$ generic mutually independent elements.

Subclaim. For any $u \in U^2$, there is a $j \in \{1, \ldots, d\}$ such that $\text{cdim}(u/a_j) = \text{cdim}(u)$.

Indeed, suppose this does not hold. By dimension calculus,

$$\text{cdim}(\langle u, a_{i+1}\rangle/\{a_1, \ldots, a_i\}) = \text{cdim}(u/\{a_1, \ldots, a_i, a_{i+1}\})$$
$$+ \text{cdim}(a_{i+1}/\{a_1, \ldots, a_i\})$$

and

$$\text{cdim}(\langle u, a_{i+1}\rangle/\{a_1, \ldots, a_i\}) = \text{cdim}(u/\{a_1, \ldots, a_i\})$$
$$+ \text{cdim}(a_{i+1}/\{a_1, \ldots, a_i, u\}).$$

By our assumptions, $\text{cdim}(a_{i+1}/\{a_1, \ldots, a_i\}) = d$ and $\text{cdim}(a_{i+1}/\{a_1, \ldots, a_i, u\}) \leq \text{cdim}(a_{i+1}/u) < d$; hence,

$$\text{cdim}(u/\{a_1, \ldots, a_i, a_{i+1}\}) \leq \text{cdim}(u/\{a_1, \ldots, a_i\}) - 1.$$

By applying this inequality for all $i$, we get a contradiction with the fact that $\operatorname{cdim}(u) \le d$, which proves the subclaim.

To prove the claim, consider an arbitrary element $g = u_1 \cdot u_2 \in G$. By the subclaim there is $a_j$, $1 \le j \le d$, such that $\langle u_1, u_2 \rangle$ and $a_j$ are independent. Then the equation $u_1 \cdot x = a_j$ has a solution $x = u_3$ in $U$, generic over $u_2$, and hence $u_2 \cdot y = u_3$ has a solution $y = v$ in $U$. Hence in $G$

$$u_1 \cdot u_2 \cdot v = a_j \qquad \text{and} \qquad u_1 \cdot u_2 = a_j \cdot v^{-1}.$$

The claim is proved.

In fact, more has been proved: we can decrease $U$ to a smaller open subset and still have the same $\{a_1, \ldots, a_d\}$ satisfying the statement of the claim.

Now notice that the $a_i \cdot U^{-1}$ can be identified with $\{a_i\} \times U$ and thus considered as manifolds.

We have natural partial bijections

$$a_i \cdot U^{-1} \to a_j \cdot U^{-1}, \qquad \langle a_i, v \rangle \mapsto \langle a_j, w \rangle \qquad \text{if} \qquad a_i \cdot w = a_j \cdot v.$$

The relation $\langle a_i, v \rangle \leftrightarrow \langle a_j, w \rangle$ defines a constructible equivalence relation $E$ on the set

$$S = \bigcup_{i=1}^{d} a_i \cdot U^{-1}.$$

By the definition of a constructible relation, the restriction of $E$ to some dense subset $S' \times S'$ of $S \times S$ is closed in the set. By decreasing $U$ to an open subset $U' \subseteq U$, we also decrease $S$ to a dense subset $S'$ and thus for some choice of $U$ we may assume that $E$ is a closed equivalence relation. In the same way, we can see that the ternary relation $P$ corresponding to the multiplication on $G$ is closed for some choice of $U$.

We define $G$ to be a manifold defined setwise as $G = S/E$ and covered by smooth subsets $a_i U^{-1}$ by construction. The graph $P$ of the multiplication is a closed relation on $G$. Thus $G$ is a Z-group.

Finally, notice that some open subset $U'$ of $U$ is embedded in $G$ as a dense subset. Indeed, there is a natural bijection between $U$ and $a_1 \cdot U^{-1}$, and on the other hand, $a_1 \cdot U^{-1}$ intersects any of the other $a_i \cdot U^{-1}$ at a dense (in both of them) subset via the correspondences $\langle a_1, v \rangle \mapsto \langle a_i, w \rangle$. $\qquad \square$

**Corollary 4.1.14.** *The group $J$ of jets on the curve $C$ at a generated by $U = H_{aa}/T$ is a connected Z-group of dimension 1.*

**Proposition 4.1.15 (Reineke's theorem).** *A one-dimensional connected Z-group $G$ is Abelian. In particular, $J$ is Abelian.*

*Proof.* We use Exercise 4.1.12. Assume towards a contradiction that $G$ is not Abelian. It follows that $C(G)$ is finite.

*Claim 1.* Every element of $G$ is of finite order.

Indeed, if there is an element $a \in G$ of infinite order, then the centraliser $C_G(a) = \{g \in G : g^{-1}ag = a\}$ is an infinite constructible subgroup and so must coincide with $G$. It follows that $a \in C(G)$ and the centre is infinite, contradicting our assumption. Claim proved.

We may now assume by (iii) that $G$ is centreless.

By part (iv), any two non-unit elements must be conjugated. It follows that all such elements are of the same order, say $p$.

*Claim 2.* $p$ is prime and $x^p = 1$ for all $x \in G$.

If $p = q \cdot r$, $q > 1$, $p > 1$, then by part (i) the map $x \mapsto x^q$ maps $G$ into the finite centre, which is just 1 by our assumptions. Claim proved.

Note that $p > 2$, for otherwise the group would be Abelian by elementary group-theoretic calculations.

Let $h$ be a non-unit element of $G$. Then so is $h^2$. Both are outside the centre, and hence by part (iv) there is $g \in G$ with $g^{-1}hg = h^2$. Because $2^{p-1} \equiv 1$ mod $p$, the non-unit element $g^{p-1} = g^{-1}$ commutes with $h$, but then $g$ must commute with $h$. This is a contradiction. $\square$

**Exercise 4.1.16.** *Deduce from Weil's theorem that*

*(i) a constructible subgroup of a Z-group is a Z-group; and*

*(ii) if $(G, *)$ is a group structure given by a constructible operation $*$ on a manifold $G$, then $G$ is a Z-group; that is, the operation is given by a closed ternary relation.*

Before we move on to obtain a Z-field, we want to transfer the (local) group structure back to $C$. We can trace the following from our construction.

**Remark 4.1.17.** An open subset $U(J)$ of $J$ is locally isomorphic to an open subset $U(C)$ $C$; more precisely, there is a finite unramified covering

$$p: U(C) \to U(J).$$

**Exercise 4.1.18.** *For $J$, if $e \in J$ is the identity element, then the structure induced by the group $J$ on $V_e$ is a subgroup of $J$.*

## 4.2  Getting a field

We start this section assuming the existence of an irreducible Zariski curve $J$ with an Abelian group structure on it.

Notice that the assumption (Amp) of Section 3.8 holds for $J$ because $J$ is definable in the original $C$ and thus there must be a finite-to-finite definable correspondence between $J$ and $C$ (which can in fact be traced effectively through our construction of $J$). So, we may assume that the $C$ of Sections 3.8 is our $J$.

The group operation on $J$ will be denoted $\oplus$. The graph of the operation is often denoted by the same sign. Notice that the graph of a binary operation on $J$ is Zariski isomorphic to $J \times J$ via the projection, and hence the graph of $\oplus$ is irreducible. We also define the operation $\ominus$ in the obvious way, and its graph is also irreducible.

As noticed in Exercise 4.1.18, $\oplus$ puts a commutative group structure on $\mathcal{V}_a$.

By the assumption (Amp), for every generic $\langle a, b \rangle \in J^2$, there is a one-dimensional, smooth, faithful family of curves on $J^2$ through $\langle a, b \rangle$. Now we can use the group operation to shift the curves point-wise

$$\langle x, y \rangle \mapsto \langle x - a,\ y - b \rangle$$

so our family becomes the family of curves through $\langle 0, 0 \rangle$. Thus we have proved the following Lemma.

**Lemma 4.2.1.** *There is a one-dimensional, smooth, faithful family $N$ of curves on $J^2$ through $\langle 0, 0 \rangle$.*

As in Section 3.8, each curve $g \in N$ defines a local bijection of $\mathcal{V}_0$ onto itself (for both $a$ and $b$ are equal to 0 now) with all the properties that we have already established.

We continue to assume that all families of curves we consider are smooth, faithful, and at least one-dimensional.

**Definition 4.2.2.** For $\tilde{g}_1, \tilde{g}_2$ branches of curves in $J^2$, we define the **sum of branches curves** $\tilde{g}_1 \oplus \tilde{g}_2$ to be

$$\{(x, y) \in J^2 : \exists z_1, z_2 \in J\ ((x, z_1) \in \tilde{g}_1\ \&\ (x, z_2) \in \tilde{g}_2\ \&\ ((z_1, z_2, y) \in \oplus))\}.$$

Equivalently, if we use notation $\tilde{g}$ for local functions on $\mathcal{V}_0$,

$$\tilde{g}_1 \oplus \tilde{g}_2(x) = \tilde{g}_1(x) \oplus \tilde{g}_2(x),$$

is a well-defined function from $\mathcal{V}_0$ into $\mathcal{V}_0$. This is also true for $\ominus$.

**Lemma 4.2.3.** *Tangency is preserved under $\oplus$ and $\ominus$. That is, if $G_1$ and $G_2$ are families of branches of curves through $\langle 0, 0 \rangle$ and if $g_1 \; T \; g_2$ and $f_1 \; T \; f_2$, then $g_1 \oplus f_1 \; T \; g_2 \oplus f_2$ and $g_1 \ominus f_1 \; T \; g_2 \ominus f_2$.*

*Proof.* Use the criterion 3.8.14(2). Given $x, y \in \mathcal{V}_0$, we want to find $g_1' \in \mathcal{V}_{g_1}$, $f_1' \in \mathcal{V}_{f_1}$, $g_2' \in \mathcal{V}_{g_2}$, and $f_2' \in \mathcal{V}_{f_2}$ such that

$$(g_1' \oplus f_1')(x) = y = (g_2' \oplus f_2')(x). \tag{4.3}$$

Choose first $y_1, y_2 \in \mathcal{V}_0$ so that $y_1 \oplus y_2 = y$ and then by tangency there exist $g_1' \in \mathcal{V}_{g_1}$, $f_1' \in \mathcal{V}_{f_1}$, $g_2' \in \mathcal{V}_{g_2}$, and $f_2' \in \mathcal{V}_{f_2}$ such that

$$g_1'(x) = y_1 = f_1'(x) \qquad \text{and} \qquad g_2'(x) = y_2 = f_2'(x).$$

By definition, Equation (4.3) follows.

The same is true for $\ominus$. $\qquad\qquad\square$

We are going to consider, as in Section 4.1, the operations $\circ$ and $^{-1}$ of composition of (branches of) curves through $\langle 0, 0 \rangle$ which again obviously give curves through $\langle 0, 0 \rangle$.

**Lemma 4.2.4.** *Let $G_1, G_2,$ and $G_3$ be families of curves through $\langle 0, 0 \rangle$ and $g_i \in G_i$, $i = 1, 2, 3$.*
*Then*

$$(g_1 \oplus g_2) \circ g_3 \; T \; (g_1 \circ g_3) \oplus (g_2 \circ g_3).$$

*Proof.* By definition of $\oplus$ on curves, we have for all $x \in \mathcal{V}_0$ and for all $g_i' \in G_i$

$$(g_1' \oplus g_2') \circ g_3'(x) = g_1'(g_3'(x)) \oplus g_2'(g_3'(x)) = (g_1' \circ g_3')(x) \oplus (g_2' \circ g_3')(x). \tag{4.4}$$

We can choose, for any $z_1, z_2 \in \mathcal{V}_0$, an element $g_i' \in \mathcal{V}_{g_i} \cap G_i$ for each $i = 1, 2, 3$ such that $g_i'(z_1) = z_2$. Using this, for any $x, y \in \mathcal{V}_0$, one can find $g_i' \in \mathcal{V}_{g_i} \cap G_i$, such that $(g_1' \oplus g_2') \circ g_3'(x) = y$. This gives Equation (4.4) and the tangency. $\qquad\square$

Obviously the symmetric distribution law holds too:

$$g_3 \circ (g_1 \oplus g_2) \; T \; (g_3 \circ g_1) \oplus (g_3 \circ g_2).$$

Thus, by applying the argument of Section 4.1 to the pair of operations we get the following corollary.

**Corollary 4.2.5.** *There is a one-dimensional irreducible manifold $U$ and constructible irreducible ternary relations $P, S \subseteq U^3$ which are the graphs of partial maps $U^2 \to U$ and determine a **partial Z-field structure** on $U$; that is,*

*P determines a pre-group structure on U with a binary operation*

$$\langle u, v \rangle \mapsto u \cdot v$$

*S determines a pre-group structure on U with a binary operation*

$$\langle u, v \rangle \mapsto u + v,$$

*and the distribution law holds for any generic triple $\langle u, v, w \rangle \in U^3$ :*

$$(u + v) \cdot w = uw + vw \text{ and } w \cdot (u + v) = wu + wv.$$

We can go from here, in analogy with Theorem 4.1.13, to construct a Z-field $K$ with a dense partial field $U' \subseteq U$ embedded in it. However, working with two partial operations at a time is not very convenient, so instead we use an algebraic trick and replace the partial field structure with a pre-group structure.

**Lemma 4.2.6.** *In the notation of Corollary 4.2.5, there is a non-commutative metabelian Z-pre-group structure $T(U)$ on the set $U \times U$.*

*For some dense open $U' \subseteq U$, there is a Z-embedding of the pre-group $T(U')$ into a connected Z-group $G$, $\dim G = 2$, and $G$ is a solvable group with finite centre.*

*Proof.* Define a partial multiplication on $U \times U$ using the formula for the triangular metabelian matrix group

$$\begin{pmatrix} u & v \\ 0 & u^{-1} \end{pmatrix},$$

that is $(u_1, v_1) * (u_2, v_2) = (u_1 u_2, \ u_1 v_2 + v_1 u_2^{-1})$. Denote $T(U)$ as the pre-group structure on $U \times U$.

By Theorem 4.1.13, there is a Z-group $G$ such that some open dense $G' \subseteq G$ is Z-isomorphic to an open dense subset of $T(U)$. We may assume that the latter is $T(U')$ for some open dense $U' \subseteq U$, and even, to simplify the notation, that $G'$ is just $T(U)$. It follows in particular that $\dim G = 2$.

By Corollary 4.1.14, $(U, \cdot)$ and $(U, +)$ are commutative pre-groups.

Moreover, $T(U)$ satisfies the metabelian identity

$$[u_1, v_1] * [u_2, v_2] = [u_2, v_2] * [u_1, v_1] \text{ for every generic } \langle u_1, v_1, u_2, v_2 \rangle \in T(U)^4,$$

where $[u, v] = u * v * u^{-1} * v^{-1}$. Indeed, one can calculate that the metabelian identity holds for generic variables using the partial field identities and commutativity.

On the other hand, the generics of $T(U)$ do not satisfy the class-2 nilpotence identity

$$u * [v, w] = [v, w] * u,$$

as the similar calculation with generic $\langle u, v, w \rangle \in T(U)^3$ shows.

We claim that the group $G$ is metabelian; that is, the metabelian identity $[u_1, v_1] * [u_2, v_2] = [u_2, v_2] * [u_1, v_1]$ holds for every $u_1, v_1, u_2, v_2 \in G$.

Indeed, using Proposition 3.5.28, we can always find in $M^*$ a generic $\langle u_1', v_1', u_2', v_2' \rangle \in G^4$ which specialises to the given $\langle u_1, v_1, u_2, v_2 \rangle$. Because the metabelian identity holds for the generic quadruple and the specialisation preserves the operation, we get the identity for $\langle u_1, v_1, u_2, v_2 \rangle$.

The centre $C(G)$ of $G$ is finite. Indeed, assuming a contradiction that $\dim C(G) \geq 1$, we have that $\dim G/C(G) \leq 1$. This means that $G/C(G)$ is a stable minimal group. By Reineke's theorem, it must be Abelian, which implies that the class-2 nilpotence identity holds for $G$, a contradiction. $\square$

**Theorem 4.2.7.** *There exists a Z-field in M.*

*Proof.* We start with the Z-group $G$ with the finite centre $C(G)$ constructed in Lemma 4.2.6.

*Claim 1.* The quotient group $G/C(G)$ is a connected Z-group.

It follows from Proposition 3.7.22 with $N = G$ (smooth by definition) and the equivalence relation

$$E(x, y) \equiv y \in x \cdot C(G).$$

Notice that such an equivalence relation is e-irreducible because $E$ has an obvious irreducible decomposition,

$$E(x, y) \equiv \bigvee_{g \in C(G)} y = xg.$$

It follows from the claim and Exercise 4.1.12(i) that the orbit of any non-unit element is infinite. In particular, $G$ and $G/C(G)$ have no finite normal subgroups and thus $G/C(G)$ has a trivial centre. So, from now on we assume that $G$ is centreless.

Consider the commutator subgroup $[G, G]$ of $G$ and $1 \neq a \in [G, G]$. Because $[G, G]$ is a normal proper subgroup of $G$, the orbit $a^G$ is a (constructible) subset of $[G, G]$ and $0 < \dim a^G < \dim G = 2$; that is, $\dim a^G = 1$. If $b$ is another non-unit element of $[G, G]$ then $a^G = b^G$ or $a^G \cap b^G = \emptyset$. Thus, we have a partition of $[G, G] \setminus \{1\}$ into one-dimensional orbits, and by considering dimensions we conclude that there are only finitely many such orbits.

Thus, $[G, G]$ is a constructible group of dimension 1, and it must contain a connected subgroup $[G, G]^0$, normal in $G$, of the same dimension. Because $[G, G]^0$ is irreducible and normal, this argument shows that

$$[G, G]^0 = a^G \cup \{1\}$$

for some non-unit element $a$.

Now denote $K^+ := [G, G]^0$ and write the group operation in $K^+$ additively, $x + y$. The group $G$ acts on $K^+$ by conjugations; write $gx$ for $g \in G$ and $x \in K^+$ instead of $g^{-1}xg$. Then $x \mapsto gx$ is an automorphism of the Z-group $K^+$.

Notice that $g$ and $g'$ induce the same action on $K^+$ if $g^{-1}g' \in C(a)$, the centraliser of $a$ in $G$, which is a normal subgroup of $G$ because it is equal to the centraliser of the normal subgroup $[G, G]^0$. Denote $K^\times$ the quotient group $G/C(a)$. We have seen that $K^\times$ acts transitively on $K^+ \setminus \{0\}$. Also $K^\times$ is a connected one-dimensional group; hence, it is commutative. Using these facts, we easily get that, for $g, g_1, g_2 \in K^\times$,

$$ga = g_1a + g_2a \Rightarrow gx = g_1x + g_2x \qquad \text{for all } x \in K^+ \setminus \{0\}$$

and

$$g_1a = -g_2a \Rightarrow g_1x = -g_2x \qquad \text{for all } x \in K^+ \setminus \{0\}.$$

So, we can identify $g \in K^\times$ with $ga \in K^+ \setminus \{0\}$ and thus transfer the multiplicative operation from $K^\times$ to $K^+$. Thus, the manifold $K^+$ gets the two operations of a field structure. It remains to note that $+$ on $K^+$ is given by a Zariski-closed ternary relation induced from the Z-group and that the multiplication is Zariski closed by the Exercise 4.1.16(ii). $\qquad \square$

We can trace the construction to see that an open subset of $J/C(G)$ can be identified with a dense subset of $K$. Taking into account the last remark of section 4.1, we notice the following.

**Remark 4.2.8.** An open subset $U(K)$ of $K$ is locally isomorphic to an open subset $U(C)$ of $C$; more precisely, there is a finite unramified covering

$$p : U(C) \to U(K).$$

**Example 4.2.9.** It may happen that the group $(J, \oplus)$ is different from $(K, +)$. For example, it can be the multiplicative group $K^\times$. It is interesting to see how the construction of 4.2.3–4.2.5 works in this case.

So, we work in $\mathcal{V}_1 \subseteq (K^\times)^*$, and let the family of curves be rather simple, for example, of the form

$$g(v) = a \cdot v + b.$$

To have $g(1) = 1$, we must put $b = 1 - a$, so

$$g_a(v) = a \cdot v + 1 - a.$$

The natural composition of curves leads straight to the multiplication:

$$g_a(v) \circ g_b(v) = g_{a \cdot b}.$$

Following our procedure, we can use the multiplication to introduce

$$
\begin{aligned}
g_a(v) \oplus g_b(v) &= (a \cdot v + 1 - a) \cdot (b \cdot v + 1 - b) \\
&= ab \cdot v^2 + (a + b - 2ab) \cdot v + ab + 1 - a - b \\
&= f(v).
\end{aligned}
$$

This curve has a derivative at 1, $f'(1) = a + b$. Thus, it is tangent to $g_{a+b} = (a + b) \cdot v + 1 - a - b$, and thus

$$g_a(v) \oplus g_b(v) \; T \; g_{a+b}.$$

## 4.3 Projective spaces over a Z-field

We assume here that $K$ is a one-dimensional, irreducible, pre-smooth Zariski structure on a field $K$ obtained by an expansion of the natural language (of Zariski-closed algebraic relations). Such a Z-structure has been constructed in Section 4.2 by means of the ambient Zariski structure M.

### 4.3.1 Projective spaces as Zariski structures

By the standard procedure, we construct projective spaces $\mathbf{P}^n(K) = \mathbf{P}^n$ over $K$ as a quotient

$$\mathbf{P}^n(K) = \left( K^{n+1} \setminus \langle 0, \dots 0 \rangle \right) / \sim$$

where

$$\langle x_0, \dots, x_n \rangle \sim \langle y_0, \dots, y_n \rangle \Leftrightarrow \exists \lambda \in K^\times : \langle x_0, \dots, x_n \rangle = \langle \lambda y_0, \dots, \lambda y_n \rangle.$$

We let $\theta_n$ stand for the natural mapping

$$\theta_n : K^{n+1} \setminus \langle 0, \dots 0 \rangle \to \mathbf{P}^n(K).$$

There is a classical presentation of $\mathbf{P}^n$ as a Z-set (of type A). Let

$$U_i = \{\langle x_0, \ldots, x_n \rangle \in K^{n+1} : x_i \neq 0\}.$$

We may identify $\theta_n(U_i)$ with

$$\tilde{U}_i = \{\langle y_0, \ldots, y_n \rangle \in K^{n+1} : y_i = 1\}$$

because in every class $\langle x_0, \ldots, x_n \rangle / \sim$ there is a unique element with $x_i = 1$. Obviously, with this identification in mind, $\mathbf{P}^n = \bigcup_{i=0}^n \tilde{U}_i$ and the conditions of Proposition 3.7.16 are satisfied; thus, $\mathbf{P}^n$ with the corresponding collection of closed subsets is a pre-smooth Zariski structure. In particular, $\theta_n$ is a Z-morphism.

### 4.3.2 Completeness

Though we cannot prove the completeness of the Zariski structure on $\mathbf{P}^n$, we prove a weaker condition that is sufficient for our purposes.

**Definition 4.3.1.** We say that the Zariski topology on a set $N$ is **weakly complete** if, given a pre-smooth $P$, a closed subset $S \subseteq P \times N$, and the projection $\mathrm{pr} : P \times N \to P$ such that the image $\mathrm{pr}\, S$ is dense in $P$, we have $\mathrm{pr}\, S = P$.

**Proposition 4.3.2.** $\mathbf{P}^n$ *is weakly complete.*

*Proof.* We are given $S \subseteq P \times \mathbf{P}^n$ such that $S$ projects onto a dense subset of pre-smooth $P$.

We may assume that $S$ is irreducible and so is $P$.

Let $\theta$ be the map from $P \times \left(K^{n+1} \setminus (0)\right)$ to $P \times \mathbf{P}^n$ given as $\theta(p, x) = (p, \theta_n x)$. Let $\tilde{S}$ be the closure in $P \times K^{n+1}$ of $\theta^{-1}(S)$. Because $\theta$ is a Z-morphism, $\theta^{-1}(S)$ is closed in $P \times \left(K^{n+1} \setminus (0)\right)$, and so

$$\tilde{S} \cap P \times \left(K^{n+1} \setminus (0)\right) = \theta^{-1}(S).$$

For $\lambda \in K$, $x \in K^{n+1}$, and $p \in P$, write $\lambda \cdot (p, x)$ for $(p, \lambda x)$. This is a Z-isomorphism of $P \times K^{n+1}$ onto itself, if $\lambda \in K^\times$.

*Claim 1.* If $(p, x) \in \tilde{S}$ and $\lambda \in K^\times$, then $\lambda \cdot (p, x) \in \tilde{S}$.

Indeed, if $(p, x) \in \theta^{-1}(S)$, then $(p, \lambda x) \in \theta^{-1}(S) \subseteq \tilde{S}$. Hence, $\theta^{-1}(S) \subseteq \lambda^{-1}\tilde{S}$, but the latter is closed as the inverse image of closed under a Z-morphism, so $\tilde{S} \subseteq \lambda^{-1}\tilde{S}$. This proves the claim.

We have $\dim \tilde{S} = \dim S + 1$. Let $Z$ be a component of $\tilde{S}$ of maximal dimension.

*Claim 2.* For any $\lambda \in K^\times$, $\lambda^{-1} \cdot Z = Z$.

Indeed, $\lambda^{-1} \cdot Z$ is also a component of maximal dimension. Thus, the group $K^\times$ acts on the finite set of components of maximal dimension. Hence,

$$\{\lambda \in K^\times : \lambda^{-1} \cdot Z = Z\}$$

is a closed subgroup of finite index, but $K$ is irreducible, and this is the whole of $K^\times$.

Because

$$\dim Z = \dim \tilde{S} = \dim S + 1 \geq \dim P + 1 > \dim(P \times (0)),$$

we have

$$\dim Z \cap \left(P \times \left(K^{n+1} \setminus (0)\right)\right) = \dim S + 1.$$

Thus, $\theta\left(Z \cap \left(P \times \left(K^{n+1} \setminus (0)\right)\right)\right)$ is dense in $S$, so it projects onto a dense subset of $P$. Let $p$ be a generic element of $P$. Then there exists $(p, x) \in Z$. Let $Z(p, K^{n+1})$ be the fibre of $Z$ over the point $p$. Because $Z$ is $K^\times$-invariant, $K^\times \cdot (p, x) \subseteq Z$. The latter is closed, and the closure of $K^\times \cdot (p, x)$ is $(p, Kx)$; thus, $(p, 0) \in Z$. Now, the closed set

$$\{p \in P : (p, 0) \in Z\}$$

contains a generic element and hence is equal to $P$. This proves that $Z(p, K^{n+1}) \neq \emptyset$ for any $p \in P$. By pre-smoothness,

$$\dim Z(p, K^{n+1}) \geq \dim Z + \dim K^{n+1} - \dim(P \times K^{n+1}) \geq 1.$$

Hence, $Z(p, K^{n+1})$ is infinite and so contains a point $(p, x)$ with $x \neq 0$. So, $(p, x) \in \tilde{S} \cap \left(P \times (K^{n+1} \setminus (0))\right) = \theta^{-1}S$, and hence $(p, \theta_n x) \in S$, showing that $p \in \text{pr } S$ for any $p \in P$. $\qquad \square$

### 4.3.3 Intersection theory in projective spaces

We continue the study of the Zariski geometry on the field $K$.

In this section, we are going to consider intersection theory for curves on $\mathbf{P}^2$, where by 'curve' we understand a constructible one-dimensional subset of $\mathbf{P}^2$, given as a member of an irreducible family $L$. We fix the notation $L^d$ for the family of curves of degree d on $\mathbf{P}^2$, which are given by obvious polynomial equations, and thus by classical facts $L^d$ can be canonically identified as projective space $\mathbf{P}^{n(d)}$ for $n(d) = (d+2)(d+1)/2 - 1$, which is also the dimension of the space.

We do not know yet whether all the curves on $\mathbf{P}^2$ are algebraic, and the crucial question is how, given a general curve $c$, an arbitrary algebraic curve $l \in L^d$ intersects $c$.

Two curves $l_1$ and $l_2$ from families $L_1$, $L_2$, respectively, are said in this section to be **simply tangent** (with respect to $L_1$ and $L_2$) at a common point $p$, if

$$\text{ind}_p\,(l_1, l_2/L_1, L_2) > 1$$

or the curves have a common infinite component (see Section 3.6.3).

In particular, we say that $c$ is simply tangent to a curve $l \in L$ at a point $p \in c$ if

$$\text{ind}_p\,(c, l/\{c\}, L) > 1,$$

that is, there is a generic $l' \in \mathcal{V}_l \cap L$ such that

$$\#(l' \cap c \cap \mathcal{V}_p) > 1.$$

We may assume that $c$ is irreducible.

Most of the time, we say *just tangent* instead of *simply tangent*.

**Lemma 4.3.3.** *There is a finite subset $c_s$ of $c$ such that for any $d > 0$ and any line $l \in L^d$ tangent to $c$ at a point $p \in c \setminus c_s$ there is a straight line $l_p \in L^1$ tangent to both $l$ and $c$.*

*Proof.* By definition of tangency, there are distinct points $p', p'' \in \mathcal{V}_p \cap l'_1 \cap l'_2$ for generic $l' \in L^d \cap \mathcal{V}_l$. Obviously, $\langle p', p'' \rangle$ is generic in $c \times c$. Take now the straight line $l'_p$ passing through $p', p''$.

The statement of the lemma is obviously true if $c$ coincides with a straight line in infinite number of points. We assume that this is not the case and so the set of straight lines intersecting $c$ in two distinct points is of dimension 2; that is, the set contains a generic straight line. It follows that $l'_p$ is a generic line in $L_1$.

*Claim.* For some finite subset $c_s$ of $c$ depending on $c$ only, for any $p \in c \setminus c_s$, there is $l_p \in L_1$ such that $l'_p \in \mathcal{V}_{l_p}$, for $l'_p$ chosen as before.

*Proof.* Let $S \subseteq c \times c \times L^1$ be the locus of $\langle p', p'', l'_p \rangle$. Because $L^1$ can be identified with $\mathbf{P}^2$, we write

$$S \subseteq c \times c \times \mathbf{P}^2.$$

The projection of $S$ on $c \times c$ is dense in $c \times c$ because $\langle p', p'' \rangle$ is generic. By removing a finite number of points, we assume that $c$ is pre-smooth. Thus, we are under assumptions of Proposition 4.3.2. Hence, $S$ projects on the whole of $c \times c$. In other words, $S$ is a covering of $c \times c$ with generic fibres consisting of one point. By Lemma 3.5.13, all but finitely many points of $c \times c$ are regular for the covering. We remove a finite subset of $c$ and may now assume that all the points of $c \times c$ are regular for $S$ and $p$ belongs to the new $c$ (or rather to $c \setminus c_s$).

There exists $l_p \in L^1$ such that $\langle p, p, l_p \rangle \in S$. By the multiplicity property, Proposition 3.6.9(ii), $l_p$ is determined uniquely by $p$. By Proposition 3.6.2, for our $\langle p', p'' \rangle$ there exists $l''_p \in \mathcal{V}_{l_p} \cap L^1$ such that $\langle p', p'', l''_p \rangle \in S$, but by the same multiplicity property, $l''_p$ is determined uniquely by $\langle p', p'' \rangle$, so $l''_p = l'_p$. Claim proved.

It is easy to see now that $l_p$ is tangent to $l$ and $c$ at $p$. Indeed, by construction,

$$\mathrm{ind}_p\big(l_p, l/L_1, L_d\big) \geq \#l'_p \cap l' \cap \mathcal{V}_p \geq 2$$

and

$$\mathrm{ind}_p\big(l_p, c/L_1, \{c\}\big) \geq \#l'_p \cap c \cap \mathcal{V}_p \geq 2.$$

$\square$

**Remark 4.3.4.** The proof also shows that for each $p \in c \setminus c_s$, there is a unique straight line $l_p$ tangent to $c$ at $p$. Correspondingly, $c_s$ may be interpreted as the set of singular points of the curve $c$.

**Lemma 4.3.5.** *Let $l_1 + \cdots + l_d$ denote a curve of degree $d$, which is a union of $d$ distinct straight lines with no three of them passing through a common point. A straight line $l$ is tangent to $l_1 + \cdots + l_d$ with respect to $L^1, L^d$ iff it coincides with one of the lines $l_1, \ldots, l_d$.*

*Proof.* If $l$ is tangent to $l_1 + \cdots + l_d$, then they intersect in less than $d$ points or have an infinite intersection. In our case, only the latter is possible. $\square$

### 4.3.4 Generalised Bezout and Chow theorems

**Definition 4.3.6.** For a family $L$ of curves called **degree of curves of** $L$, the number

$$\deg(L) = \mathrm{ind}(L, L^1),$$

that is, the number of points in the intersection of a generic member of $L$ with a generic straight line.

For algebraic curves $a$ of (usual) degree $d$, we always assume $a \in L^d$ and write $\deg(a)$ instead of $\deg(L^d)$ (which is just $d$, of course).

For a single curve $c$, we write $\deg^*(c)$ for $\deg(\{c\})$, that is, for the number of points in the intersection of $c$ with a generic straight line.

**Theorem 4.3.7 (generalised Bezout theorem).** *For any curve $c$ on* $\mathbf{P}^2$,

$$\mathrm{ind}(\{c\}, L^d) = d \cdot \deg^* c;$$

*in particular, for an algebraic curve $a$,*

$$\#c \cap a \leq \deg^* c \cdot \deg a.$$

*Proof.* Assume $a \in L^d$ and take $l_1 + \cdots + l_d$ such that none of the straight lines is tangent to $c$ [use Proposition 3.6.15(iii) to find such lines].

*Claim.* The $c$ and $l_1 + \cdots + l_d$ are not tangent.

By Lemma 4.3.3, the tangency would imply that there is an $l$ tangent to $c$ and tangent to $l_1 + \cdots + l_d$. Lemma 4.3.5 says this is not the case.

The claim implies that the intersection indices of the curves $c$ and $l_1 + \cdots + l_d$ are equal to 1 for any point in the intersection, so by formula 3.6.15(ii)

$$\mathrm{ind}(\{c\}, L^d) = \#c \cap (l_1 + \cdots + l_d) = d \cdot \deg^* c.$$

On the other hand,

$$\#c \cap a \leq \mathrm{ind}(\{c\}, L^d)$$

because point multiplicities are minimal for generic intersections, by Proposition 3.6.15(iii).  □

**Lemma 4.3.8.** *If a curve $c$ is a subset of an algebraic curve $a$, then $c$ is algebraic.*

*Proof.* There is a birational map of $a$ into an algebraic group $J(a)$ (the Jacobian of $a$ or the multiplicative group of the field in the case when $a$ is a rational curve), which is Abelian and divisible. So, we assume $a \subseteq J(a)$. The properties of this embedding imply that for $g = \dim J(a)$ for any generic $x \in J(a)$, there is unique, up to the order, representation $x = y_1 + \cdots + y_g$ for some $y_1, \ldots, y_g$ from $a$. Now, if $c$ is a proper subset of $a$, then the set $a \setminus c$ is also of dimension 1 and so

$$\{y_1 + \cdots + y_g : y_1, \ldots, y_g \in c\} \quad \text{and} \quad \{y_1 + \cdots + y_g : y_1, \ldots, y_g \in a \setminus c\}$$

are disjoint subsets of $J(a)$ of the same dimension (equal to Morley rank) $g$, and this implies $J(a)$ is of Morley degree greater than 1. Consequently, the group has a proper subgroup of finite index [the connected component; see Poizat's work (1987)], contradicting divisibility of $J(a)$. $\qquad\square$

**Theorem 4.3.9 (generalised Chow theorem).** *Any closed subset of* $\mathbf{P}^n$ *is an algebraic sub-variety of* $\mathbf{P}^n$.

*Proof.* First we prove the statement for $n = 2$. Let $c$ be a closed subset of $\mathbf{P}^2$. W.l.o.g. we may assume $c$ is an irreducible curve. Let $q = \deg^* c$. Now choose $d$ such that $(d - 1)/2 > q$. Fix a subset $X$ of $c$, containing exactly $d \cdot q + 1$ points. Then by dimension considerations there is a curve $a \in L^d$ containing $X$. By the generalised Bezout theorem, $\#(c \cap a) \le d \cdot q$ or the intersection is infinite. Because the former is excluded by construction, $c$ has an infinite intersection with the algebraic curve $a$. Thus $c$ coincides with an irreducible component of $a$, which is also algebraic by Lemma 4.3.8.

Now we consider a closed subset $Q \subseteq \mathbf{P}^n$ and assume that for $\mathbf{P}^{n-1}$ the statement of the theorem is true. By fixing a generic subspace $H \subset \mathbf{P}^n$ isomorphic to $\mathbf{P}^{n-2}$ and a generic straight line $l \subset \mathbf{P}^n$, we can fibre $\mathbf{P}^n$ and $Q$ by linear subspaces $S_p$, generated by $H$ and a point $p$, varying in $l$. Evidently, $S_p$ is biregularly isomorphic to $\mathbf{P}^{n-1}$, and we can apply the inductive hypothesis to $Q \cap S_p = Q_p$. This gives us a representation of $Q_p$ by a set of polynomial equations $f_{p,1} = 0, \ldots, f_{p,k_p} = 0$. We now consider only generic $p \in l$; thus $k_p = k$ and degrees of the polynomials do not depend on $p$.

Denote the $i$-th coefficient of the polynomial $f_{p,m}$ as $a_{i,m}(p)$. This defines on an open domain $U \subseteq \mathbf{P}^1$ a mapping $U \to K$, which corresponds to a closed curve in $\mathbf{P}^2$, and the curve is algebraic. This implies that the dependence on $p$ in the coefficients for $p \in U$ is algebraic. This allows us to rewrite the polynomials $f_{p,m}(x)$, with $x$ varying over an open subset of $S_p$, as $f'_m(p, x)$ where now $\langle p, x \rangle$ varies over an open subset of $\mathbf{P}^n$. Thus, $q$ coincides with an algebraic closed set on a subset, open in both of them, and so $q$ coincides with the algebraic closed set. $\qquad\square$

**Theorem 4.3.10 (purity theorem).** *Any relation $R$ induced on $K$ from $M$ is definable in the natural language and so is constructible.*

*Proof.* By elimination of quantifiers for Zariski structures, it suffices to prove the statement for closed $R \subseteq K^n$. Consider the canonical (algebraic) embedding of $K^n$ into $\mathbf{P}^n$ and the closure $\bar{R} \subseteq \mathbf{P}^m$ of $R$. By the generalised Chow theorem, $\bar{R}$ is an algebraic subset of $\mathbf{P}^n$, but $R = \bar{R} \cap K^n$. $\qquad\square$

## 4.4 The classification theorem

### 4.4.1 Main theorem

**Theorem 4.4.1.** *Let* M *be a Zariski structure satisfying (EU) and* $C$ *be a pre-smooth Zariski curve in* M. *Assume that* $C$ *is non-linear (equivalently* $C$ *is ample in the sense of Section 3.8). Then there is a non-constant continuous map*

$$f : C \to \mathbf{P}^1(K).$$

*Moreover,* $f$ *is a finite map [* $f^{-1}(x)$ *is finite for every* $x \in C$ *], and for any* $n$, *for any definable subset* $S \subseteq C^n$, *the image* $f(S)$ *is a constructible subset (in the sense of algebraic geometry) of* $[\mathbf{P}^1(K)]^n$.

*Proof.* The field $K$ has been constructed in Section 4.2, Theorem 4.2.7. By the construction, $K$ is definable in terms of the structure on $C$ (induced from M); more precisely, $K$ is a 1-manifold with respect to $C$. So, there is a finite-to-finite closed relation $F \subseteq C \times K$ which projects on an open (co-finite) subset of $D \subseteq C$ and an open subset $R \subseteq K$.

*Claim.* There exist a co-finite subset $D' \subseteq C$ and a non-constant continuous function $s : D' \to K$.

*Proof.* Consider $x \in D$ and let $F(x, K)$ be the fibre over $x$ of the covering $\langle x, y \rangle \mapsto x$ of $D$. Assuming that $x$ is generic in $D$, there is an $n$ such that $F(x, K) = \{y_1, \ldots, y_n\}$, with $y_i \neq y_j$, for any $i < j \leq n$. Let $s_1, \ldots, s_n$ be the standard symmetric functions of $n$-variables:

$$s_1(\bar{y}) = y_1 + \cdots + y_n, \ s_2(\bar{y}) = y_1 \cdot y_2 + \cdots y_1 \cdot y_n + \cdots + y_{n-1} \cdot y_n, \ldots,$$
$$\ldots s_n(\bar{y}) = y_1 \cdot y_2 \cdots y_n.$$

We can identify each $s_i(\bar{y})$ as a function of the unordered set $F(x, K)$. Because it is a function of $x$, write it $s_i(x)$. Conversely, by elementary algebra, the set $\{y_1, \ldots, y_n\}$ is exactly the set of all roots of the polynomial $p_x(v) = v^n + s_1(x)v^{n-1} + \cdots + s_n(x)$. Hence, on a co-finite subset $D'$ of $C$, we have defined functions

$$s_i : D' \to K, \ i = 1, \ldots, n.$$

At least one of the functions, say $s_i$, must be non-constant and have a co-finite image in $K$, because $\{y \in K : \exists x \in D' \ F(x, y)\}$ is co-finite in $K$. Because the graph of the function $s_i$ is constructible, by possibly decreasing the domain of the function by a finite subset, we can get the condition that the graph of $s_i$ is closed in $D' \times K$. This means that $s_i$ is continuous on $D'$. Claim proved.

Now consider a continuous function $s : D' \to K$ of the claim and the closure $S \subseteq C \times \mathbf{P}^1(K)$ of its graph in $C \times \mathbf{P}^1(K)$. $S$ is irreducible because the graph of $s$ is irreducible. By Proposition 4.3.2, $S$ is a covering of $C$. Also, the covering $S$ is finite of multiplicity 1 in generic points. Then by Corollary 3.5.14, $S$ is finite in every point of $C$. By multiplicity properties [3.6.9(ii)], $S$ has multiplicity 1 in every point; that is, $S$ is the graph of a function called again $f$. □

### 4.4.2 Meromorphic functions on a Zariski set

**Definition 4.4.2.** For a given Zariski set $N$ and a field $K$, a continuous function $g : N \to K$ with the domain containing an open subset of $N$ is called **Z-meromorphic on $N$**.

Notice that the sum and the product of two meromorphic functions on $N$ are Z-meromorphic. Moreover, if $g$ is Z-meromorphic and non-zero, then $1/g$ is a meromorphic function. In other words, the set of meromorphic functions on $N$ forms a field.

We denote $K_Z(N)$ as the **field of Z-meromorphic functions on $N$**.

**Remark 4.4.3.** Notice that if the characteristic of $K$ is $p > 0$, then with any Z-meromorphic function $f$ one can associate distinct Z-meromorphic functions $\phi^n \circ f$, $n \in \mathbb{Z}$, where $\phi$ is the Frobenius automorphism of the field $x \mapsto x^p$.

Of course, for negative $n$ the map $\phi^n : K \to K$ is not rational. So, when $N$ is an algebraic curve, $K_Z(N)$ is the inseparable closure of the field $K(N)$ of rational functions on $N$, that is, the closure of $K(N)$ under the powers of the Frobenius.

**Proposition 4.4.4 (second part of the main theorem).** *Under the assumptions of Theorem 4.4.1, there exists a smooth algebraic quasi-projective curve $X$ over $K$ and a Zariski epimorphism*

$$\psi : C \to X$$

*with the universality property: for any algebraic curve $Y$ over $K$ and a Zariski epimorphism $\tau : C \to Y$, there exists a Zariski epimorphism $\sigma : X \to Y$ such that $\sigma \circ \tau = \psi$.*

*The field $K(X)$ of rational functions is isomorphic over $K$ to a sub-field of $K_Z(C)$, and $K_Z(C)$ is equal to the inseparable closure of the field $K(X)$.*

*Proof.* We start with

*Claim 1.* tr.d.$(K_Z(C)/K) = 1$.

*Proof.* Let $g$ be a non-constant meromorphic function and $h$ an arbitrary non-constant meromorphic function defined in $M$. Choose a generic (over $M$) point $x \in C$ and let $y = g(x)$, $z = h(x)$. We have cdim$(y/M, x) = 0 =$ cdim$(x/M, y)$ and cdim$(z/M, x) = 0$. Hence, cdim$(z/M, y) = 0$. This means that there is an $M$-definable binary relation $R$ on $K$ such that $R(y, z)$ holds and $R(y, K)$ is finite. By the purity theorem, Theorem 4.3.10, $R$ is given by a polynomial equation $r(y, z) = 0$ over $K$. Because $r$, $g$ and $h$ are continuous and $x$ is generic, $r(g(v), h(v)) = 0$ for every $v \in C$. In other words, $h$ is in the algebraic closure (in the field-theoretic sense) of $g$ and $K$, for every $h \in K_Z(C)$. Claim proved.

Let $x$ again be generic in $C$ over $M$, and let $g_1, \ldots, g_n$ be non-constant Z-meromorphic functions over $K$, $y_i = g_i(x)$, $i = 1, \ldots, n$. By dimension calculations, $g_i^{-1}(y_i)$ is finite, so there exists an $n$ such that

$$[x] = \bigcap_{i=1}^{n} g_i^{-1}(y_i)$$

is minimal possible. It implies that for any other meromorphic function $h$, the value $y = h(x)$ is determined by the class $[x]$ (and $h$). Consequently, $y \in$ dcl$(y_1, \ldots, y_n, K)$, the definable closure of $y_1, \ldots, y_n$, $K$. This means that there is a definable (constructible) relation $H(v_1, \ldots, v_n, v)$ over $K$ such that $H(y_1, \ldots, y_n, v)$ is satisfied by the unique element $y$. It follows that there is a partial Zariski-continuous function $p$ such that $y = p(y_1, \ldots, y_n)$. It follows that $h(v) = p(g_1(v), \ldots, g_n(v))$ for all $v \in C$.

*Claim 2.* The only constructible functions on an algebraically closed field are of the form $\phi^n \circ r$, $n \in \mathbb{Z}$, where $r$ is rational and $\phi$ is the Frobenius.

*Proof.* A function $f(\bar{v})$ defined over a sub-field $K_0$ determines, for every $\bar{v}$, the unique element $w = f(\bar{v})$. Assume $\bar{v}$ is generic over $K_0$. By Galois theory, $w$ is in the inseparable closure of $K_0(\bar{v})$; that is, $w = \phi^n(r(\bar{v}))$ for some $r(\bar{v}) \in K_0(\bar{v})$.

Hence we have proved the next claim.

*Claim 3.* $K_Z(C)$ is equal to the inseparable closure of the field $K(g_1, \ldots, g_n)$. More precisely, every element $f \in K_Z(C)$ is of the form $\phi^m(g)$, for some $g \in K(g_1, \ldots, g_n)$ and a $m \in \mathbb{Z}$.

Let $X$ be the image in $[\mathbf{P}^1(K)]^n$ of $C$ under the map

$$\psi : v \mapsto \langle g_1(v), \ldots, g_n(v) \rangle.$$

By the purity theorem, this is a constructible set. By assumptions, $X$ is one-dimensional and irreducible, so it has the form $X = \bar{X} \setminus X_0$, for some closed (projective) curve $\bar{X}$ and a finite subset $X_0$. This is, by definition [see e.g. Shafarevich's work (1977)], a quasi-projective algebraic curve. By construction, $X$ is locally isomorphic to $C$; hence, by Proposition 3.6.26, $X$ is pre-smooth. By the analysis of Proposition 3.5.9, $X$ is Zariski isomorphic, via a map $e$, to a smooth algebraic curve. We may assume this curve is $X$ by applying to $\psi$ the Zariski isomorphism $e$. Also, the domain of $\psi$ must be $C$ because the space $[\mathbf{P}^1(K)]^n$ is weakly complete in our Zariski topology.

Obviously, to every rational function $f : X \mapsto K$, we can put in correspondence the unique Z-meromorphic function $\psi^*(f)$ on $C : v \mapsto f(\psi(v))$. Let $K(X)$ be the field of rational functions on $X$. Then $\psi^*$ embeds $K(X)$ into $K_Z(C)$, and the co-ordinate functions of $X$ correspond to $g_1, \ldots, g_n$.

Suppose now $\tau : C \to Y$ is a continuous epimorphism on an algebraic curve $Y$ over $K$. Then $\tau^*$ embeds the field $K(Y)$ of rational functions on $Y$ into $K_Z(C)$. That is, by claim 3, $K(Y) \subseteq \phi^m(K(X))$ for some $m \in \mathbb{Z}$. This embedding can be represented as $\sigma_0^* : K(Y) \to \phi^m(K(X))$ for some rational epimorphism $\sigma_0 : \phi^{-m}(X) \to Y$ of algebraic curves. This finally gives the Zariski epimorphism $\sigma = \sigma_0 \circ \phi^m$, sending $X$ onto $Y$. $\qquad\Box$

**Remark 4.4.5.** In general, $\psi$ is not a bijection; that is, $C$ is not isomorphic to an algebraic curve. See Section 5.1 for examples.

### 4.4.3 Simple Zariski groups are algebraic

The main theorem is crucial to prove the algebraicity conjecture for groups definable in pre-smooth Zariski structures.

**Theorem 4.4.6.** *Let $G$ be a simple Zariski group satisfying (EU) such that some one-dimensional irreducible Z-subset $C$ in $G$ is pre-smooth. Then $G$ is Zariski isomorphic to an algebraic group $\hat{G}(K)$, for some algebraically closed field $K$.*

*Proof.* We start with a general statement.

*Claim 1.* Let $G$ be a simple group of finite Morley rank. Then $\mathrm{Th}(G)$ is categorical in uncountable cardinals (in the language of groups). Moreover, $G$ is almost strongly minimal.

This is a direct consequence of the indecomposability theorem on finite Morley rank groups and is proved by Poizat (1987), Proposition 2.12.

*Claim 2.* Given a strongly minimal set $C$ definable in $G$, there is a definable relation $F \subseteq G \times C^m$, $m = \mathrm{rk}\, G$, establishing a finite-to-finite correspondence between a subset $R \subseteq G$ and a subset $D \subseteq C^m$ such that $\dim(G \setminus R) < m$ and $\dim(C^m \setminus D) < m$.

This is a consequence of the proof of the previous statement.

*Claim 3.* For $G$ as in the condition of the theorem, there exists a non-constant meromorphic function $G \to K$.

To prove the claim, first notice that $C$ in claim 2 can be replaced by $K$ because there is a finite-to-finite correspondence between the two. Now apply the argument with symmetric functions as in the proof of the claim in the main theorem. This proves the present claim.

Now consider the field $K_Z(G)$ of meromorphic functions $G \to K$. Each $g \in G$ acts on $K_Z(G)$ by $f(x) \mapsto f(g \cdot x)$. This gives a representation of $G$ as the group of automorphisms of $K_Z(G)$. This action can also be seen as the $K$-linear action on the $K$-vector space $K_Z(G)$. As is standard in the theory of algebraic groups (Rosenlicht's theorem), by using the purity theorem one can see that there is a $G$-invariant finite-dimensional $K$-subspace $V$ of $K_Z(G)$. Hence, $G$ can be represented as a definable subgroup $\hat{G}(K)$ of $\mathrm{GL}(V)$, and by the purity theorem again this subgroup is algebraic. This representation is an isomorphism because $G$ is simple. □

Notice that pre-smoothness is paramount for this proof. In the case of Zariski groups without pre-smoothness (which, of course, still are of finite Morley rank by Theorem 3.2.8), the algebraicity conjecture remains open.

# 5

# Non-classical Zariski geometries

## 5.1 Non-algebraic Zariski geometries

**Theorem 5.1.1.** *There exists an irreducible pre-smooth Zariski structure (in particular of dimension 1) which is not interpretable in an algebraically closed field.*

### The construction

Let $\mathbf{M} = (M, \mathcal{C})$ be an irreducible pre-smooth Zariski structure, $G \leq \mathrm{ZAut}\, M$ (Zariski-continuous bijections), acting freely on $M$ and for some $\tilde{G}$ with **finite** $H$:

$$1 \to H \to \tilde{G} \to^{\mathrm{pr}} G \to 1.$$

Consider a set $X \subseteq M$ of representatives of $G$-orbits: for each $a \in M$, $G \cdot a \cap X$ is a singleton.

Consider the formal set

$$\tilde{M}(\tilde{G}) = \tilde{M} = \tilde{G} \times X$$

and the projection map

$$\mathbf{p} : (g, x) \mapsto \mathrm{pr}\,(g) \cdot x.$$

Consider also, for each $f \in \tilde{G}$, the function

$$f : (g, x) \mapsto (fg, x).$$

We thus have obtained the structure

$$\tilde{\mathbf{M}} = (\tilde{M},\ \{f\}_{f \in \tilde{G}} \cup \mathbf{p}^{-1}(\mathcal{C}))$$

on the set $\tilde{M}$ with relations induced from M together with maps $\{f\}_{f \in \tilde{G}}$. We set the closed subsets of $\tilde{M}^n$ to be exactly those which are definable by positive quantifier-free formulas with parameters. Obviously, the structure M and the map $\mathbf{p} : \tilde{M} \to M$ are definable in $\tilde{M}$ because, for each $f \in \tilde{G}$,

$$\forall v \; \mathbf{p} f(v) = f \mathbf{p}(v),$$

the image $\mathbf{p}(S)$ of a closed subset $S \subseteq \tilde{M}^n$ is closed in M. We define dim $S :=$ dim $\mathbf{p}(S)$.

**Lemma 5.1.2.** *The isomorphism type of $\tilde{M}$ is determined by M and $\tilde{G}$ only. The theory of $\tilde{M}$ has quantifier elimination. $\tilde{M}$ is an irreducible pre-smooth Zariski structure.*

*Proof.* One can use obvious automorphisms of the structure to prove quantifier elimination. The statement of the claim then follows by checking the definitions. The detailed proof is given by Hrushovski and Zilber (1993), Proposition 10.1.                                                                                          □

**Lemma 5.1.3.** *Suppose H does not split; that is, for every proper $G_0 < \tilde{G}$*

$$G_0 \cdot H \neq \tilde{G}.$$

*Then, every equidimensional Zariski expansion $\tilde{M}'$ of $\tilde{M}$ is irreducible.*

*Proof.* Let $C = \tilde{M}'$ be an $|H|$-cover of the variety $M$, so dim $C =$ dim $M$ and $C$ has at most $|H|$ distinct irreducible components, say $C_i$, $1 \leq i \leq n$. For generic $y \in M$, the fibre $\mathbf{p}^{-1}(y)$ intersects every $C_i$ [otherwise $\mathbf{p}^{-1}(M)$ is not equal to $C$].

Hence, $H$ acts transitively on the set of irreducible components. $\tilde{G}$ acts transitively on the set of irreducible components, so the setwise stabiliser $G^0$ of $C_1$ in $\tilde{G}$ is of index $n$ in $\tilde{G}$ and also $H \cap \tilde{G}^0$ is of index $n$ in $H$. Hence,

$$\tilde{G} = G^0 \cdot H, \qquad \text{with} \qquad H \nsubseteq G^0,$$

contradicting our assumptions.                                                                                    □

**Lemma 5.1.4.** *$\tilde{G} \leq \mathrm{ZAut}\,\tilde{M}$; that is, $\tilde{G}$ is a subgroup of the group of Zariski-continuous bijections of $\tilde{M}$.*

*Proof.* Immediate by construction.                                                                          □

**Lemma 5.1.5.** *Suppose M is a rational or elliptic curve (over an algebraically closed field K of characteristic zero), H does not split, $\tilde{G}$ is nilpotent, and for some big integer $\mu$ there is a non-Abelian subgroup $G_0 \leq \tilde{G}$:*

$$|\tilde{G} : G_0| \geq \mu.$$

*Then $\tilde{M}$ is not interpretable in an algebraically closed field.*

*Proof.* First we show the following:

*Claim.* Without loss of generality we may assume that $\tilde{G}$ is infinite.

Recall that $G$ is a subgroup of the group ZAut $M$ of rational (Zariski) automorphisms of $M$. Every algebraic curve is birationally equivalent to a smooth one, so $G$ embeds into the group of birational transformations of a smooth rational curve or an elliptic curve. Now remember that any birational transformation of a smooth algebraic curve is biregular. If $M$ is rational, then the group ZAut $M$ is PGL(2, $K$). Choose a semi-simple (diagonal) $s \in$ PGL(2, $K$) to be an automorphism of infinite order such that $\langle s \rangle \cap G = 1$ and $G$ commutes with $s$. Then we can replace $G$ by $G' = \langle G, s \rangle$ and $\tilde{G}$ by $\tilde{G}' = \langle \tilde{G}, s \rangle$ with the trivial action of $s$ on $H$. One can easily see from the construction that the $\tilde{M}'$ corresponding to $\tilde{G}'$ is the same as $\tilde{M}$, except for the new definable bijection corresponding to $s$.

We can use the same argument when $M$ is an elliptic curve, in which case the group of automorphisms of the curve is given as a semi-direct product of a finitely generated Abelian group (complex multiplication) acting on the group on the elliptic curve $E(K)$.

Now, assuming that $\tilde{M}$ is definable in an algebraically closed field $K'$, we have that $K$ is definable in $K'$. It is known to imply that $K'$ is definably isomorphic to $K$, so we may assume that $K' = K$.

Also, because dim $\tilde{M}$ = dim $M = 1$, it follows that $\tilde{M}$ up to finitely many points is in a bijective definable correspondence with a smooth algebraic curve, say $C = C(K)$.

By the previous argument, $\tilde{G}$ then is embedded into the group of rational automorphisms of $C$.

The automorphism group is finite if genus of the curve is two or more, so by the claim we can have only a rational or elliptic curve for $C$.

Consider first the case when $C$ is rational. The automorphism group then is PGL(2, $K$). Because $\tilde{G}$ is nilpotent, its Zariski closure in PGL(2, $K$) is an infinite nilpotent group $U$. Let $U^0$ be the connected component of $U$, which is a normal subgroup of finite index. By a theorem of A. I. Malcev, there is a number $\mu$ (dependent only on the size of the matrix group in question but not on $U$) such that some normal subgroup $U^0$ of $U$ of index at most $\mu$ is a subgroup of the unipotent group

$$\begin{pmatrix} 1 & z \\ 0 & 1 \end{pmatrix}.$$

This is Abelian, contradicting the assumption that $\tilde{G}$ has no Abelian subgroups of index less than $\mu$.

In case $C$ is an elliptic curve, the group of automorphisms is a semi-direct product of a finitely generated Abelian group (complex multiplication) acting freely on the Abelian group of the elliptic curve. This group has no nilpotent non-Abelian subgroups. This finishes the proof of the lemma and of the theorem.                                                                    □

In general, it is harder to analyse the situation when $\dim M > 1$ because the group of birational automorphisms is not so immediately reducible to the group of biregular automorphisms of a smooth variety in higher dimensions. Nevertheless, the same method can prove the useful fact that the construction produces examples essentially of non-algebra-geometric nature.

**Proposition 5.1.6.**

(i) *Suppose $M$ is an Abelian variety, $H$ does not split, and $\tilde{G}$ is nilpotent, not Abelian. Then $\tilde{M}$ cannot be an algebraic variety with $p : \tilde{M} \to M$ a regular map.*

(ii) *Suppose $M$ is the (semi-Abelian) variety $(K^{\times})^n$. Suppose also that $\tilde{G}$ is nilpotent and for some big integer $\mu = \mu(n)$ has no Abelian subgroup $G_0$ of index bigger than $\mu$. Then $\tilde{M}$ cannot be an algebraic variety with $p : \tilde{M} \to M$ a regular map.*

*Proof.*

(i) If $M$ is an Abelian variety and $\tilde{M}$ were algebraic, the map $p : \tilde{M} \to M$ has to be unramified because all its fibers are of the same order (equal to $|H|$). Hence $\tilde{M}$, being a finite unramified cover, must have the same universal cover $M$ has. So, $\tilde{M}$ must be an Abelian variety as well. The group of automorphisms of an Abelian variety $\mathcal{A}$ without complex multiplication is the Abelian group $\mathcal{A}(K)$, a contradiction.

(ii) The same argument as in (i) proves that $\tilde{M}$ has to be isomorphic to $(K^{\times})^n$. The Malcev theorem cited previously finishes the proof.                    □

**Proposition 5.1.7.** *Suppose $M$ is a $K$-variety and, in the construction of $\tilde{M}$, the group $\tilde{G}$ is finite. Then $\tilde{M}$ is definable in any expansion of the field $K$ by a total linear order.*

*In particular, if $M$ is a complex variety, $\tilde{M}$ is definable in the reals.*

*Proof.* Extend the ordering of $K$ to a linear order of $M$, and define

$$S := \{s \in M : s = \min G \cdot s\}.$$

The rest of the construction of $\tilde{M}$ is definable.                                              □

**Remark 5.1.8.** In other known examples of non-algebraic $\tilde{M}$ (with $G$ infinite), $\tilde{M}$ is still definable in any expansion of the field $K$ by a total linear order. In particular, see the example considered in the next section.

## 5.2 Case study

### 5.2.1 The $N$-cover of the affine line

We assume here that the characteristic of $K$ is 0.

Let $a, b \in K$ be additively independent.

$G$ acts on $K$:

$$ux = a + x, \quad vx = b + x.$$

Taking $M$ to be $K$, this determines, by Sub-section 5.1, a pre-smooth non-algebraic Zariski curve $\tilde{M}$ which from now on we denote $\mathbf{P}_N$, and $P_N$ stands for the universe of this structure.

The corresponding definition for the covering map $\mathbf{p} : \tilde{M} \to M = K$ then gives us

$$\mathbf{p}(ut) = a + \mathbf{p}(t), \qquad \mathbf{p}(vt) = b + \mathbf{p}(t). \tag{5.1}$$

### 5.2.2 Semi-definable functions on $\mathbf{P}_N$

**Lemma 5.2.1.** *There are functions $y$ and $z$*

$$P_N \to K$$

*satisfying the following* **functional equations,** *for any $t \in P_N$:*

$$\mathbf{y}^N(t) = 1, \qquad \mathbf{y}(ut) = \epsilon \mathbf{y}(t), \qquad \mathbf{y}(vt) = \mathbf{y}(t) \tag{5.2}$$
$$\mathbf{z}^N(t) = 1, \qquad \mathbf{z}(ut) = \mathbf{z}(t), \qquad \mathbf{z}(vt) = \mathbf{y}(t)^{-1} \cdot \mathbf{z}(t). \tag{5.3}$$

*Proof.* Choose a subset $S \subseteq M = K$ of representatives of $G$-orbits; that is, $K = G + S$. By the construction in Section 5.1, we can identify $P_N = \tilde{M}$ with $\tilde{G} \times S$ in such a way that $\mathbf{p}(\gamma s) = \mathrm{pr}(\gamma) + s$. This means that, for any $s \in S$ and $t \in \tilde{G} \cdot s$ of the form $t = \mathbf{u}^m \mathbf{v}^n [\mathbf{u}, \mathbf{v}]^l \cdot s$,

$$\mathbf{p}(\mathbf{u}^m \mathbf{v}^n [\mathbf{u}, \mathbf{v}]^l \cdot s) := ma + nb + s,$$

set also

$$\mathbf{y}(\mathbf{u}^m \mathbf{v}^n [\mathbf{u}, \mathbf{v}]^l \cdot s) := \epsilon^m$$
$$\mathbf{z}(\mathbf{u}^m \mathbf{v}^n [\mathbf{u}, \mathbf{v}]^l \cdot s) := \epsilon^l.$$

This satisfies Equations (5.2) and (5.3). $\qquad \square$

**Remark 5.2.2.** Notice that these follow from Equations (5.1)–(5.3):

1. **p** is surjective and $N$-to-1, with fibres of the form

$$\mathbf{p}^{-1}(\lambda) = Ht, \qquad H = \{[\mathbf{u}, \mathbf{v}]^l : 0 \le l < N\}.$$

2. $\mathbf{y}([\mathbf{u}, \mathbf{v}]t) = \mathbf{y}(t)$, and
3. $\mathbf{z}([\mathbf{u}, \mathbf{v}]t) = \epsilon \mathbf{z}(t)$.

**Definition 5.2.3.** Define the **band function** on $K$ as a function $\mathrm{bd} : K \to K[N]$.

Set for $\lambda \in K$

$$\mathrm{bd}(\lambda) = \mathbf{y}(t), \qquad \text{if } \mathbf{p}(t) = \lambda.$$

This is well defined by the previous remark.

By acting by **u** on $t$ and using Equations (5.1) and (5.2), we have

$$\mathrm{bd}(a + \lambda) = \epsilon \mathrm{bd}\,\lambda. \tag{5.4}$$

By acting by **v**, we obtain

$$\mathrm{bd}(b + \lambda) = \mathrm{bd}\,\lambda. \tag{5.5}$$

**Lemma 5.2.4.** *The structure* $\mathbf{P}_N$ *is definable in*

$$(K, +, \cdot, \mathrm{bd}).$$

*Proof.* Set

$$P_N = K \times K[N] = \{\langle x, \epsilon^l \rangle : x \in K,\ l = 0, \ldots, N-1\}$$

and define the maps

$$\mathbf{p}(\langle x, \epsilon^l \rangle) := x, \quad \mathbf{y}(\langle x, \epsilon^l \rangle) := \mathrm{bd}(x), \qquad \mathbf{z}(\langle x, \epsilon^l \rangle) := \epsilon^{-l}.$$

Also define

$$\mathbf{u}(\langle x, \epsilon^l \rangle) := \langle a + x, \epsilon^l \rangle), \qquad \mathbf{v}(\langle x, \epsilon^l \rangle) := \langle b + x, \epsilon^l \mathrm{bd}(x) \rangle.$$

One checks easily that the action of $\tilde{G}$ is well defined and that Equations (5.1)–(5.3) hold. $\qquad \square$

Assuming that $K = \mathbb{C}$ and for simplicity that $a \in i\mathbb{R}$ and $b \in \mathbb{R}$, both nonzero, we may define, for $z \in \mathbb{C}$,

$$\mathrm{bd}(z) := \exp\left(\frac{2\pi i}{N}\left[\mathrm{Re}\left(\frac{z}{a}\right)\right]\right).$$

This satisfies Equations (5.4) and (5.5) and so $P_N$ over $\mathbb{C}$ is definable in $\mathbb{C}$ equipped with the previous measurable but not continuous function.

### 5.2.3 Space of semi-definable functions

Let $\mathcal{H}$ be the $K$-algebra of semi-definable functions on $P_N$ generated by $x$, $y$, $z$. We define linear operators $\mathbf{X}$, $\mathbf{Y}$, $\mathbf{Z}$, $\mathbf{U}$, and $\mathbf{V}$ on $\mathcal{H}$:

$$\begin{aligned}
\mathbf{X} &: \psi(t) \mapsto \mathbf{p}(t) \cdot \psi(t), \\
\mathbf{Y} &: \psi(t) \mapsto \mathbf{y}(t) \cdot \psi(t), \\
\mathbf{Z} &: \psi(t) \mapsto \mathbf{z}(t) \cdot \psi(t), \\
\mathbf{U} &: \psi(t) \mapsto \psi(\mathbf{u}t), \\
\mathbf{V} &: \psi(t) \mapsto \psi(\mathbf{v}t).
\end{aligned} \tag{5.6}$$

Denote the group generated by the operators $\mathbf{U}$, $\mathbf{V}$, $\mathbf{U}^{-1}$, and $\mathbf{V}^{-1}$ as $\tilde{G}^*$; denote the $K$-algebra $K[\mathbf{X}, \mathbf{Y}, \mathbf{Z}]$ as $\mathfrak{X}_\epsilon$ (or simply $\mathfrak{X}$); and denote the extension of the $K$ algebra $\mathfrak{X}_\epsilon$ by $\tilde{G}^*$ as $\mathcal{A}_\epsilon$ (or simply $\mathcal{A}$).

$\mathcal{H}$ with the action of $\mathcal{A}$ on it is determined uniquely up to isomorphism by the defining Equations (5.1)–(5.3) and so is independent of the arbitrariness in the choices of $\mathbf{x}$, $\mathbf{y}$, and $\mathbf{z}$. The algebra $\mathcal{A}_\epsilon$ is determined by its generators and the following relations, with $\mathbf{E}$ standing for the commutator $[\mathbf{U}, \mathbf{V}]$:

$$\begin{aligned}
& \mathbf{XY} = \mathbf{YX}; \mathbf{XZ} = \mathbf{ZX}; \mathbf{YZ} = \mathbf{ZY}; \\
& \mathbf{Y}^N = 1; \mathbf{Z}^N = 1; \\
& \mathbf{UX} - \mathbf{XU} = a\mathbf{U}; \mathbf{VX} - \mathbf{XV} = b\mathbf{V}; \\
& \mathbf{UY} = \epsilon\mathbf{YU}; \mathbf{YV} = \mathbf{VY}; \\
& \mathbf{ZU} = \mathbf{UZ}; \\
& \mathbf{VZ} = \mathbf{YZV}; \\
& \mathbf{UE} = \mathbf{EU}; \mathbf{VE} = \mathbf{EV}; \mathbf{E}^N = 1.
\end{aligned} \tag{5.7}$$

### 5.2.4 Representation of $\mathcal{A}$

Let $\mathrm{Max}(\mathfrak{X})$ be the set of isomorphism classes of one-dimensional irreducible $\mathfrak{X}$-modules.

**Lemma 5.2.5.** $\mathrm{Max}(\mathfrak{X})$ *can be represented by one-dimensional modules* $\langle e_{\mu,\xi,\zeta} \rangle$ $(= K e_{\mu,\xi,\zeta})$ *for* $\mu \in K, \xi, \zeta \in K[N]$, *defined by the action on the generating vector as follows:*

$$\mathbf{X} e_{\mu,\xi,\zeta} = \mu e_{\mu,\xi,\zeta}, \qquad \mathbf{Y} e_{\mu,\xi,\zeta} = \xi e_{\mu,\xi,\zeta}, \qquad \mathbf{Z} e_{\mu,\xi,\zeta} = \zeta e_{\mu,\xi,\zeta}.$$

*Proof.* This is a standard fact of commutative algebra.                    □

Assuming $K$ is endowed with the function $\mathrm{bd} : K \to K[N]$, we call $\langle \mu, \xi, \zeta \rangle$ **real oriented** if

$$\mathrm{bd}\mu = \xi.$$

Correspondingly, we call the module $\langle e_{\mu,\xi,\zeta} \rangle$ real oriented if $\langle \mu, \xi, \zeta \rangle$ is real oriented.

$\mathrm{Max}^+(\mathfrak{X})$ denotes the subspace of $\mathrm{Max}(\mathfrak{X})$ consisting of real-oriented modules $\langle e_{\mu,\xi,\zeta} \rangle$.                    □

**Lemma 5.2.6.** *$\langle \mu, \xi, \zeta \rangle$ is real oriented if and only if*

$$\langle \mu, \xi, \zeta \rangle = \langle \mathbf{p}(t), \mathbf{y}(t), \mathbf{z}(t) \rangle$$

*for some $t \in T$.*

*Proof.* It follows from the definition of bd that $\langle \mathbf{p}(t), \mathbf{y}(t), \mathbf{z}(t) \rangle$ is real oriented.

Assume now that $\langle \mu, \xi, \zeta \rangle$ is real oriented. Because $\mathbf{p}$ is a surjection, there is $t' \in T$ such that $\mathbf{p}(t') = \mu$. By the definition of bd, $\mathbf{y}(t') = \mathrm{bd}\,\mu$. By the remark after Lemma 5.2.1, both values stay the same if we replace $t'$ with $t = [\mathbf{u}, \mathbf{v}]^k\, t'$. By the same remark, for some $k$, $\mathbf{z}(t) = \zeta$.                    □

Now we introduce an infinite-dimensional $\mathcal{A}$-module $\mathcal{H}_0$. As a vector space, $\mathcal{H}_0$ is spanned by $\{ e_{\mu,\xi,\zeta} : \mu \in K, \xi, \zeta \in K[N] \}$. The action of the generators of $\mathcal{A}$ on $\mathcal{H}_0$ is defined on $e_{\mu,\xi,\zeta}$ in accordance with the defining relations of $\mathcal{A}$. So, because

$$\mathbf{XU}e_{\mu,\xi,\zeta} = (\mathbf{UX} - a\mathbf{U})e_{\mu,\xi,\zeta} = (\mu - a)\mathbf{U}e_{\mu,\xi,\zeta},$$

$$\mathbf{YU}e_{\mu,\xi,\zeta} = \epsilon^{-1}\mathbf{UY}e_{\mu,\xi,\zeta} = \epsilon^{-1}\xi\mathbf{U}e_{\mu,\xi,\zeta},$$

$$\mathbf{ZU}e_{\mu,\xi,\zeta} = \mathbf{UZ}e_{\mu,\xi,\zeta} = \zeta\mathbf{U}e_{\mu,\xi,\zeta},$$

and

$$\mathbf{XV}e_{\mu,\xi,\zeta} = (\mathbf{VX} - b\mathbf{V})e_{\mu,\xi,\zeta} = (\mu - b)\mathbf{V}e_{\mu,\xi,\zeta},$$

$$\mathbf{YV}e_{\mu,\xi,\zeta} = \mathbf{VY}e_{\mu,\xi,\zeta} = \xi\mathbf{U}e_{\mu,\xi,\zeta},$$

$$\mathbf{ZV}e_{\mu,\xi,\zeta} = \mathbf{VY}^{-1}\mathbf{Z}e_{\mu,\xi,\zeta} = \xi^{-1}\zeta\mathbf{V}e_{\mu,\xi,\zeta},$$

we set

$$\mathbf{U}e_{\mu,\xi,\zeta} := e_{\mathbf{u}\langle\mu,\xi,\zeta\rangle}, \text{ with } \mathbf{u}\langle\mu, \xi, \zeta\rangle = \langle\mu - a, \epsilon^{-1}\xi, \zeta\rangle$$

and

$$\mathbf{V}e_{\mu,\xi,\zeta} := e_{\mathbf{v}\langle\mu,\xi,\zeta\rangle}, \qquad \text{with } \mathbf{v}\langle\mu,\xi,\zeta\rangle = \langle\mu - b, \xi, \xi^{-1}\zeta\rangle.$$

We may now identify $\text{Max}(\mathfrak{X})$ as the family of one-dimensional $\mathfrak{X}$-eigenspaces of $\mathcal{H}_0$. Correspondingly, we call the $\mathfrak{X}$-module (state) $\langle e_{\mu,\xi}\rangle$ real oriented if $\langle\mu,\xi\rangle$ is real oriented. $\mathcal{H}_0^+$ denotes the linear subspace of $\mathcal{H}_0$ spanned by the positively oriented states $\langle e_{\mu,\xi}\rangle$. We denote $\text{Max}^+(\mathfrak{X})$ as the family of one-dimensional real-oriented $\mathfrak{X}$-eigenspaces of $\mathcal{H}_0$, or *states*, as such things are referred to in physics literature.

**Theorem 5.2.7.**

(i) *There is a bijective correspondence $\Xi : \text{Max}^+(\mathfrak{X}) \to P_N$ between the set of real-oriented $\mathfrak{X}$-eigensubspaces of $\mathcal{H}_0$ and $P_N$.*

(ii) *The action of $\tilde{G}^*$ on $\mathcal{H}_0$ induces an action on $\text{Max}(\mathcal{H})$ and leaves $\text{Max}^+(\mathfrak{X})$ setwise invariant. The correspondence $\Xi$ transfers anti-isomorphically the natural action of $\tilde{G}^*$ on $\text{Max}^\omega(\mathfrak{X})$ to a natural action of $\tilde{G}$ on $P_N$.*

(iii) *The map*

$$\mathbf{p}_{\mathfrak{X}} : \langle e_{\mu,\xi,\zeta}\rangle \mapsto \mu$$

*is an $N$-to-1-surjection $\text{Max}^+(\mathfrak{X}) \to K$ such that*

$$\left(\text{Max}^+(\mathfrak{X}), \mathbf{U}, \mathbf{V}, \mathbf{p}_{\mathfrak{X}}, K\right) \cong_\xi (P_N, \mathbf{u}, \mathbf{v}, \mathbf{p}, K).$$

*Proof.*

(i) This is immediate by Lemma 5.2.6.

(ii) Indeed, by the previous definition, the action of $\mathbf{U}$ and $\mathbf{V}$ corresponds to the action on real-oriented $N$-tuples:

$$\mathbf{U} : \langle\mathbf{p}(t), \mathbf{y}(t), \mathbf{z}(t)\rangle \mapsto \langle\mathbf{p}(t) - a, \epsilon^{-1}\mathbf{y}(t),$$
$$\mathbf{z}(t)\rangle = \langle\mathbf{p}(\mathbf{u}^{-1}t), \mathbf{y}(\mathbf{u}^{-1}t), \mathbf{z}(\mathbf{u}^{-1}t)\rangle,$$
$$\mathbf{V} : \langle\mathbf{p}(t) - b, \mathbf{y}(t), \mathbf{y}(t)^{-1}\mathbf{z}(t)\rangle \mapsto \langle\mathbf{p}(\mathbf{v}^{-1}t), \mathbf{y}(\mathbf{v}^{-1}t), \mathbf{z}(\mathbf{v}^{-1}t)\rangle.$$

(iii) This is immediate from (i) and (ii). $\qquad\qquad\square$

## $C^*$-Representation

Our aim again is to introduce an involution on $\mathcal{A}$. We assume $K = \mathbb{C}$, $a = 2\pi i/N$, and $b \in \mathbb{R}$ and start by extending the space $\mathcal{H}$ of semi-definable functions with a function $\mathbf{w} : P_N \to \mathbb{C}$ such that

$$\exp \mathbf{w} = \mathbf{y}, \qquad \mathbf{w}(\mathbf{u}t) = \frac{2\pi i}{N} + \mathbf{w}(t), \qquad \mathbf{w}(\mathbf{v}t) = \mathbf{w}(t).$$

We can easily do this by setting, as in Equation (5.2.1),

$$\mathbf{w}(\mathbf{u}^m \mathbf{v}^n [\mathbf{u}, \mathbf{v}]^l \cdot s) := \frac{2\pi i m}{N}.$$

Now we extend $\mathcal{A}$ to $\mathcal{A}^{\#}$ by adding the new operator

$$\mathbf{W} : \psi \mapsto \mathbf{w}\psi$$

which obviously satisfies

$$\mathbf{WX} = \mathbf{XW}, \qquad \mathbf{WY} = \mathbf{YW}, \qquad \mathbf{WZ} = \mathbf{ZW}.$$

$$\mathbf{UW} = \frac{2\pi i}{N} + \mathbf{WU}, \qquad \mathbf{VW} = \mathbf{WV}.$$

We set

$$\mathbf{U}^* := \mathbf{U}^{-1}, \qquad \mathbf{V}^* := \mathbf{V}^{-1}$$

$$\mathbf{Y}^* := \mathbf{Y}^{-1}, \qquad \mathbf{W}^* := -\mathbf{W}, \qquad \mathbf{X}^* := \mathbf{X} - 2\mathbf{W},$$

implying that $\mathbf{U}$, $\mathbf{V}$, and $\mathbf{Y}$ are unitary and $i\mathbf{W}$ and $\mathbf{X} - \mathbf{W}$ are formally self-adjoint.

**Proposition 5.2.8.** *There is a representation of $\mathcal{A}^{\#}$ in an inner product space such that $\mathbf{U}$, $\mathbf{V}$, and $\mathbf{Y}$ act as unitary and $i\mathbf{W}$ and $\mathbf{X} - \mathbf{W}$ as self-adjoint operators.*

*Proof.* Let $\mathcal{H}_R$ be the subspace of the inner product space $\mathcal{H}_0$ spanned by vectors $e_{\mu, \xi, \zeta}$ such that

$$\mu = x + \frac{2\pi i k}{N}, \qquad \xi = e^{\frac{2\pi i k}{N}}, \zeta = e^{\frac{2\pi i m}{N}}, \qquad \text{for } x \in \mathbb{R}, \ k, m \in \mathbb{Z}. \quad (5.8)$$

One checks that $\mathcal{H}_R$ is closed under the action of $\mathcal{A}$ on $\mathcal{H}_0$ defined in Lemma 5.2.4, that is, $\mathcal{H}_R$ is an $\mathcal{A}$-sub-module. We also define the action by $\mathbf{W}$

$$\mathbf{W} : e_{\mu, \xi, \zeta} \mapsto \frac{2\pi i k}{N} e_{\mu, \xi, \zeta}$$

for $\mu = x + (2\pi i k / N)$. This obviously agrees with the defining relations of $\mathcal{A}^{\#}$, so $\mathcal{H}_R$ is an $\mathcal{A}^{\#}$-sub-module of $\mathcal{H}_0$.

Now $\mathbf{U}$ and $\mathbf{V}$ are unitary operators on $\mathcal{H}_R$ because they transform the orthonormal basis into itself. $\mathbf{Y}$ is unitary because its eigenvectors form the orthonormal basis and the corresponding eigenvalues are of absolute value 1. $i\mathbf{W}$ and $\mathbf{X} - \mathbf{W}$ are self-adjoint because their eigenvalues on the orthonormal basis are the reals $-(2\pi k/N)$ and $x$. $\qquad \square$

**Comments**

1. Note that following Theorem 5.2.7, we can treat the set of 'states' $\langle e_{\mu,\xi,\zeta}\rangle$ satisfying Equation (5.8) as a sub-structure of $P_N$. When one applies the definition of the band function 5.2.3 to these, one gets

$$\operatorname{bd}\mu = \exp\frac{2\pi ik}{N}, \qquad \text{for} \qquad \mu = x + \frac{2\pi ik}{N}.$$

In other words, in this representation the band function is again a way to separate the real and imaginary parts of the complex numbers involved.

2. The discrete nature of the imaginary part of $\mu$ in Equation (5.8) is necessitated by two conditions: the interpretation of $*$ as taking adjoints and the non-continuous form of the band function. The first condition is crucial for any physical interpretation, and the second one follows from the description of the Zariski structure $P_N$. Comparing this to the real differentiable structure $P_\infty$ constructed in Section 5.2.5 as the limit of the $P_N$, we suggest interpreting the latter along with its representation via $\mathcal{A}$ in this section as the *quantisation* of the former.

### 5.2.5 Metric limit

Our aim in this section is to find an interpretation of the limit, as $N$ tends to $\infty$, of structures $P_N$ in 'classical' terms. Classical here means 'using function and relations given in terms of real manifolds and analytic functions'. Of course, we have to define the meaning of the 'limit' first. We found a satisfactory solution to this problem in the case of $P_N$ which is presented later.

### The Heisenberg Group

First, we establish a connection of the group $\tilde{G}_N$ with the *integer Heisenberg group* $H(\mathbb{Z})$ which is the group of matrices of the form

$$\begin{pmatrix} 1 & k & m \\ 0 & 1 & l \\ 0 & 0 & 1 \end{pmatrix} \tag{5.9}$$

with $k, l, m \in \mathbb{Z}$. More precisely, $\tilde{G}_N$ is isomorphic to the group

$$H(\mathbb{Z})_N = H(\mathbb{Z})/N.Z,$$

where $N.Z$ is the central subgroup

$$N.Z = \left\{ \begin{pmatrix} 1 & 0 & Nm \\ 0 & 1 & 0 \\ 0 & 0 & 1 \end{pmatrix} : m \in \mathbb{Z} \right\}.$$

Similarly the *real Heisenberg group* $H(\mathbb{R})$ is defined as the group of matrices of the form (5.9) with $k, l, m \in \mathbb{R}$. The analogue (or the limit case) of $H(\mathbb{Z})_N$ is the factor-group

$$H(\mathbb{R})_\infty := H(\mathbb{R}) \Big/ \begin{pmatrix} 1 & 0 & \mathbb{Z} \\ 0 & 1 & 0 \\ 0 & 0 & 1 \end{pmatrix}.$$

In fact, there is the natural group embedding

$$i_N : \begin{pmatrix} 1 & k & m \\ 0 & 1 & l \\ 0 & 0 & 1 \end{pmatrix} \mapsto \begin{pmatrix} 1 & \frac{k}{\sqrt{N}} & \frac{m}{N} \\ 0 & 1 & \frac{l}{\sqrt{N}} \\ 0 & 0 & 1 \end{pmatrix},$$

inducing the embedding $H(\mathbb{Z})_N \subset H(\mathbb{R})_\infty$.

Notice the following lemma.

**Lemma 5.2.9.** *Given the embedding $i_N$ for every $\langle u, v, w \rangle \in H(\mathbb{R})_\infty$, there is $\langle k/\sqrt{N}, l/\sqrt{N}, m/N \rangle \in i_N(H(\mathbb{Z})_N)$ such that*

$$|u - \frac{k}{\sqrt{N}}| + |v - \frac{l}{\sqrt{N}}| + |w - \frac{m}{N}| < \frac{3}{\sqrt{N}}.$$

In other words, the distance (given by the sum of absolute values) between any point of $H(\mathbb{R})_\infty$ and the set $i_N(H(\mathbb{Z})_N)$ is at most $3/\sqrt{N}$. Obviously, also the distance between any point of $i_N(H(\mathbb{Z})_N)$ and the set $H(\mathbb{R})_\infty$ is 0, because of the embedding. In other words, this defines that the **Hausdorff distance between the two sets is at most** $3/\sqrt{N}$.

In situations when the pointwise distance between sets $M_1$ and $M_2$ is defined, we also say that the Hausdorff distance between two $L$-structures on $M_1$ and $M_2$ is at most $\alpha$ if the Hausdorff distance between the universes $M_1$ and $M_2$ as well as between $R(M_1)$ and $R(M_2)$ for any $L$-predicate or graph of an $L$-operation $R$ is at most $\alpha$.

Finally, we say that an $L$-structure $M$ is the **Hausdorff limit** of $L$-structures $M_N$, $N \in \mathbb{N}$, if for each positive $\alpha$ there is $N_0$ such that for all $N > N_0$ the distance between $M_N$ and $M$ is at most $\alpha$.

**Remark 5.2.10.** It makes sense to consider the similar notion of Gromov–Hausdorff distance and Gromov–Hausdorff limit.

**Lemma 5.2.11.** *The group structure $H(\mathbb{R})_\infty$ is the Hausdorff limit of its sub-structures $H(\mathbb{Z})_N$, where the distance is defined by the embeddings $i_N$.*

*Proof.* Lemma 5.2.9 proves that the universe of $H(\mathbb{R})_\infty$ is the limit of the corresponding sequence. Because the group operation is continuous in the topology determined by the distance, the graphs of the group operations converge as well. □

### The action

Given non-zero real numbers $a, b, c$, the integer Heisenberg group $H(\mathbb{Z})$ acts on $\mathbb{R}^3$ as follows:

$$\langle k, l, m \rangle \langle x, y, s \rangle = \langle x + ak, y + bl, s + acky + abcm \rangle, \tag{5.10}$$

where $\langle k, l, m \rangle$ is the matrix (5.9).

We can also define the action of $H(\mathbb{Z})$ on $\mathbb{C} \times S^1$, equivalently on $\mathbb{R} \times \mathbb{R} \times \mathbb{R}/\mathbb{Z}$, as follows

$$\langle k, l, m \rangle \langle x, y, \exp 2\pi i s \rangle = \langle x + ak, y + bl, \exp 2\pi i (s + acky + abcm) \rangle, \tag{5.11}$$

where $x, y, s \in \mathbb{R}$.

In the discrete version intended to model Lemma 5.2.1, we consider $q/N$, $q \in \mathbb{Z}$, in place of $s \in \mathbb{R}$ and take $a = b = 1/\sqrt{N}$. We replace Equation (5.11) with

$$\langle k, l, m \rangle \langle x, y, e^{\frac{2\pi i q}{N}} \rangle = \langle x + \frac{k}{\sqrt{N}}, \ y + \frac{l}{\sqrt{N}}, \ \exp 2\pi i \frac{q + k[y\sqrt{N}] + m}{N} \rangle. \tag{5.12}$$

One can easily check that this is still an action.

Moreover, we may take $m$ modulo $N$ in Equation (5.12); that is, $\langle k, l, m \rangle \in H(\mathbb{Z})_N$, and simple calculations similar to the previous ones show the following.

**Lemma 5.2.12.** *The formula (5.12) defines the free action of $H(\mathbb{Z})_N$ on $\mathbb{R} \times \mathbb{R} \times \exp \frac{2\pi i}{N} \mathbb{Z}$ (equivalently on $\mathbb{C} \times \exp \frac{2\pi i}{N} \mathbb{Z}$).*

We think of $\langle x, y, \exp \frac{2\pi i q}{N} \rangle$ as an element $t$ of $P_N$ (see Lemma 5.2.1) and $x + iy$ as $p(t) \in \mathbb{C}$. The actions $x + iy \mapsto a + x + iy$ and $x + iy \mapsto x + i(y + b)$ are obvious rational automorphisms of the affine line $\mathbb{C}$.

We interpret the action of $\langle 1, 0, 0 \rangle$ and $\langle 0, 1, 0 \rangle$ by Equation (5.12) on $\mathbb{C} \times \exp \frac{2\pi i}{N} \mathbb{Z}$ as **u** and **v** respectively. Then the commutator [**u**, **v**] corresponds to $\langle 0, 0, -1 \rangle$, which is the generating element of the centre of $H(\mathbb{Z})_N$. In other words, the subgroup $\mathrm{gp}(\mathbf{u}, \mathbf{v})$ of $H(\mathbb{Z})_N$ generated by the two elements is isomorphic to $\tilde{G}$. Using Lemma 5.1.2, we thus get the following.

**Lemma 5.2.13.** *Under the previous assumption and notation, the structure on $\mathbb{C} \times \exp \frac{2\pi i}{N} \mathbb{Z}$ described by Equation (5.12) in the language of Sub-section 5.2.1 is isomorphic to the $\mathrm{P}_N$ of Sub-section 5.2.1 with $K = \mathbb{C}$.*

We identify $\mathrm{P}_N$ with the previous structure based on $\mathbb{C} \times \{\exp \frac{2\pi i}{N} \mathbb{Z}\}$.

Note that every group word in $\mathbf{u}$ and $\mathbf{v}$ gives rise to a definable map in $\mathrm{P}_N$. We introduce a uniform notation for such definable functions.

Let $\alpha$ be a monotone non-decreasing converging sequence of the form

$$\alpha = \left\{ \frac{k_N}{\sqrt{N}} : k_N, N \in \mathbb{Z}, \ N > 0 \right\}.$$

We call such a sequence **admissible** if there is an $r \in \mathbb{R}$ such that

$$\left| r - \frac{k_N}{\sqrt{N}} \right| \leq \frac{1}{\sqrt{N}}. \tag{5.13}$$

Given $r \in \mathbb{R}$ and $N \in \mathbb{N}$, one can easily find $k_N$ satisfying Equation (5.13) and so construct an $\alpha$ converging to $r$, which we denote $\hat{\alpha}$:

$$\hat{\alpha} := \lim \alpha = \lim_N \frac{k_N}{\sqrt{N}}.$$

We denote the set of all admissible sequences converging to a real on $[0, 1]$ as $I$, so

$$\{\hat{\alpha} : \alpha \in I\} = \mathbb{R} \cap [0, 1].$$

For each $\alpha \in I$, we introduce two operation symbols $\mathbf{u}_\alpha$ and $\mathbf{v}_\alpha$. We use $\mathrm{P}_N^\#$ to denote the definable expansion of $\mathrm{P}_N$ by all such symbols with the interpretation

$$\mathbf{u}_\alpha = \mathbf{u}^{k_N}, \qquad \mathbf{v}_\alpha = \mathbf{v}^{k_N} \quad (k_N\text{-multiple of the operation}),$$

if $k_N/\sqrt{N}$ stands in the $N$th position in the sequence $\alpha$.

Note that the sequence

$$dt := \left\{ \frac{1}{\sqrt{N}} : N \in \mathbb{N} \right\}$$

is in $I$ and $\mathbf{u}_{dt} = \mathbf{u}$, $\mathbf{v}_{dt} = \mathbf{v}$ in all $\mathrm{P}_N^\#$.

We now define the structure $\mathrm{P}_\infty$ to be the structure on sorts $\mathbb{C} \times S^1$ (denoted $P_\infty$) and sort $\mathbb{C}$, with the field structure on $\mathbb{C}$ and the projection map $\mathbf{p} :$

$\langle x, y, e^{2\pi i s}\rangle \mapsto \langle x, y\rangle \in \mathbb{C}$, and definable maps $\mathbf{u}_\alpha$ and $\mathbf{v}_\beta$, $\alpha, \beta \in I$, acting on $\mathbb{C} \times S^1$ [in accordance with the action by $H(\mathbb{R})_\infty$] as follows:

$$\mathbf{u}_\alpha(\langle x, y, e^{2\pi i s}\rangle) = \langle \hat{\alpha}, 0, 0\rangle\langle x, y, e^{2\pi i s}\rangle = \langle x + \hat{\alpha}, \; y, \; e^{2\pi i(s + \hat{\alpha}y)}\rangle$$
$$\mathbf{v}_\beta(\langle x, y, e^{2\pi i s}\rangle) = \langle 0, \hat{\beta}, 0\rangle\langle x, y, e^{2\pi i s}\rangle = \langle x, \; y + \hat{\beta}, \; e^{2\pi i s}\rangle \tag{5.14}$$

**Theorem 5.2.14.** $P_\infty$ *is the Hausdorff limit of structures* $P_N^\#$.

*Proof.* The sort $\mathbb{C}$ is the same in all structures.

The sort $P_\infty$ is the limit of its sub-structures $P_N$ because $S^1$ ($= \exp i\mathbb{R}$) is the limit of $\exp \frac{2\pi i}{N}\mathbb{Z}$ in the standard metric of $\mathbb{C}$. Also, the graph of the projection map $\mathbf{p} : P_\infty \to \mathbb{C}$ is the limit of $\mathbf{p} : P_N \to \mathbb{C}$ for the same reason.

Finally, it remains to check that the graphs of $\mathbf{u}$ and $\mathbf{v}$ in $P_\infty$ are the limits of those in $P_N$. It is enough to see that for any $\langle x, y, \exp \frac{2\pi i q}{N}\rangle \in P_N$ the result of the action by $\mathbf{u}_\alpha$ and $\mathbf{v}_\beta$ calculated in $P_N^\#$ is at most at the distance $2/\sqrt{N}$ from the ones calculated in $P_\infty$, for any $\langle x, y, \exp \frac{2\pi i q}{N}\rangle \in P_\infty$. And indeed, the action in $P_N^\#$ by definition is

$$\mathbf{u}_\alpha : \langle x, y, \exp \tfrac{2\pi i q}{N}\rangle \mapsto \langle x + \tfrac{k_N}{\sqrt{N}}, \; y, \; \exp \tfrac{2\pi i}{N}(q + k_N[y\sqrt{N}])\rangle$$
$$\mathbf{v}_\beta : \langle x, y, \exp \tfrac{2\pi i q}{N}\rangle \mapsto \langle x, \; y + \tfrac{l_N}{\sqrt{N}}, \; \exp 2\pi i \tfrac{q}{N}\rangle \tag{5.15}$$

Obviously,

$$\left|\frac{k_N y}{\sqrt{N}} - \frac{k_N[y\sqrt{N}]}{N}\right| = \frac{k_N}{\sqrt{N}}\left|\frac{y\sqrt{N} - [y\sqrt{N}]}{\sqrt{N}}\right| < \frac{k_N}{\sqrt{N}}\frac{1}{\sqrt{N}} \le \frac{1}{\sqrt{N}},$$

which together with Equation (5.13) proves that the right-hand side of Equation (5.15) is at the distance at most $2/\sqrt{N}$ from the right-hand side of Equation (5.14) uniformly on the point $\langle x, y, \exp \frac{2\pi i q}{N}\rangle$. $\qquad\square$

## Comment

The structure $P_\infty$ can be seen as the principal bundle over $\mathbb{R} \times \mathbb{R}$ with the structure group $U(1)$ (the rotations of $S^1$) and the projection map $\mathbf{p}$. The action by the Heisenberg group allows us to define a *connection* on the bundle. A connection determines 'a smooth transition from a point in a fibre to a point in a nearby fibre'. As noted previously, $\mathbf{u}$ and $\mathbf{v}$ in the limit process correspond to infinitesimal actions (in a non-standard model of $P_\infty$) which can be written in the form

$$\mathbf{u}(\langle x, y, e^{2\pi i s}\rangle) = \langle x + dt, \; y, \; e^{2\pi i(s + y\,dt)}\rangle$$
$$\mathbf{v}(\langle x, y, e^{2\pi i s}\rangle) = \langle x, \; y + dt, \; e^{2\pi i s}\rangle.$$

These formulas allow us to calculate the derivative of a section

$$\psi : \langle x, y\rangle \mapsto \langle x, y, e^{2\pi i s(x,y)}\rangle$$

of the bundle in any direction on $\mathbb{R} \times \mathbb{R}$. In general, by moving infinitesimally from the point $\langle x, y \rangle$ along $x$ we get $\langle x + dt, y, \exp 2\pi i(s + ds) \rangle$. We compare this to *the parallel transport along x* given by the previous formulas, $\langle x + dt, y, \exp 2\pi i(s + y dt) \rangle$. The difference is

$$\langle 0, 0, \exp 2\pi i(s + ds) - \exp 2\pi i(s + y dt) \rangle.$$

Using the usual laws of differentiation, for the third term one gets

$$
\begin{aligned}
\exp 2\pi i(s + ds) &- \exp 2\pi i(s + y \, dt) \\
&= (\exp 2\pi i(s + ds) - \exp 2\pi i s) - (\exp 2\pi i(s + y dt) - \exp 2\pi i s) \\
&= d \exp 2\pi i s - 2\pi i y \exp 2\pi i s \, dt \\
&= \left( \frac{d \exp 2\pi i s}{dt} - 2\pi i y \exp 2\pi i s \right) dt
\end{aligned}
$$

which for a section $\psi = \exp 2\pi i s$ gives the following *covariant derivative* along $x$,

$$\nabla_x \psi = \frac{d}{dx} \psi - 2\pi i y \psi.$$

Similarly, $\nabla_y$ the covariant derivative along $y$ is just $\frac{d}{dy} \psi$, with the second term zero.

The *curvature* of the connection is by definition the commutator

$$[\nabla_x, \nabla_y] = 2\pi i.$$

In physicists' terms, this is a $U(1)$-gauge field theory over $\mathbb{R}^2$ with a constant non-zero curvature.

## 5.3 From quantum algebras to Zariski structures

In the previous section, we started with an existing construction of a series of non-classical Zariski structures and showed that, on the one hand, this series approximates in some precise sense a classical albeit non-algebraic structure and, on the other hand, each of the Zariski structures has an adequate representation by an appropriate non-commutative $C^*$-algebra. Here, in contrast, we present a construction which for any of a wide variety of $K$-algebras $\mathcal{A}$ produces in a canonical way a Zariski geometry $\tilde{V}(\mathcal{A})$ so that $\mathcal{A}$ can be seen as a (in general non-commutative) co-ordinate algebra of the structure $\tilde{V}(\mathcal{A})$. For commutative $\mathcal{A}$, the geometry $\tilde{V}(\mathcal{A})$ is just the algebraic variety corresponding to the coordinate algebra $\mathcal{A}$, and for almost all non-commutative algebras, $\tilde{V}(\mathcal{A})$ is a non-classical (that is, not definable in terms of the field $K$) Zariski structure.

Let us say a few words about the different ways, here and in Section 5.1, of representing algebras $\mathcal{A}$. Recall that the points of the structure $P_N$ associated to $\mathcal{A}$ in Section 5.1 correspond to irreducible modules of a specific commutative sub-algebra $\mathfrak{X}$ of $\mathcal{A}$, with $\mathfrak{X}$ invariant under conjugation by invertible elements of $\mathcal{A}$. The conjugation then induces definable bijections on $P_N$. In the present section, we assume that $\mathcal{A}$ has a large central sub-algebra $\mathcal{Z}$, which plays the role of $\mathfrak{X}$, and the irreducible $\mathcal{Z}$-modules, seen as points, form a classical part V of $\tilde{V}(\mathcal{A})$. In fact, V is simply Max $\mathcal{Z}$. In each point $m$ of V, we 'insert' the structure of the corresponding $\mathcal{A}$-module $M_m$ and so the universe of $\tilde{V}(\mathcal{A})$ is the union of all the modules. Note that by our assumptions all irreducible modules are finite dimensional as $K$-vector spaces. Once we have finite-dimensional modules $M_1$ and $M_2$ in our structure, we can *definably* introduce $M_1 \oplus M_2$, $M_1 \otimes M_2$, and eventually any finite dimensional module can be definably described in terms of irreducible ones. For this reason, $\tilde{V}(\mathcal{A})$ is in fact definably equivalent to the category $\mathcal{A}$-mod of all finite-dimensional $\mathcal{A}$-modules. We do not prove and neither do we use this fact, but it is a conceptually important point of the construction and an important *link to the category theory approach to geometry*. The fact that here $\mathcal{A}$ is a quantum algebra at roots of unity is important to our construction and especially to the $\mathcal{A}$-category representation. Note that $\mathcal{A}$ of Section 5.1 does not satisfy this assumption.

In more detail, we consider $K$-algebras $\mathcal{A}$ over an algebraically closed field $K$. Our assumptions imply that a typical irreducible $\mathcal{A}$-module is of finite dimension over $K$.

We introduce the structure associated with $\mathcal{A}$ as a two-sorted structure $(\tilde{V}, K)$ where $K$ is given with the usual field structure and $\tilde{V}$ is the bundle over an affine variety V of $\mathcal{A}$-modules of a fixed finite $K$-dimension $N$. Again by the assumptions, the isomorphism types of $N$-dimensional $\mathcal{A}$-modules are determined by points in V. By inserting a module $M_m$ of the corresponding type in each point $m$ of V, we get

$$\tilde{V} = \coprod_{m \in V} M_m.$$

In fact, all the modules in our case are assumed to be irreducible, but in a more general treatment in a previous work (Zilber 2008a), we only assume that $M_m$ is irreducible for any $m$ belonging to an open subset of V.

Our language contains a function symbol $\mathbf{U}_i$ acting on each $M_m$ (and so on the sort $\tilde{V}$) for each generator $\mathbf{U}_i$ of the algebra $\mathcal{A}$. We also have the binary function symbol for the action of $K$ by scalar multiplication on the modules. Because $M_m$ may be considered an $\mathcal{A}/\mathrm{Ann}\, M_m$-module, we have the bundle of finite-dimensional algebras $\mathcal{A}/\mathrm{Ann}\, M_m$, $m \in V$, represented in $\tilde{V}$. In typical cases, the intersection of all such annihilators is 0. As a consequence of this, the algebra $\mathcal{A}$ is faithfully represented by its action on the bundle of modules. This is one more reason to believe that our structure represents the category of all finite dimensional $\mathcal{A}$-modules.

We write down our description of $\tilde{V}$ as the set of first-order axioms Th($\mathcal{A}$-mod).

We prove two main theorems.

**Theorem A (5.3.5 and 5.3.10).** *The theory* Th($\mathcal{A}$-mod) *is categorical in uncountable cardinals and model complete.*

**Theorem B (5.3.11).** $\tilde{V}$ *is a Zariski geometry in both sorts.*

Theorem A is rather easy to prove, and in fact the proof does not use all of the assumptions on $\mathcal{A}$ we assumed. Yet despite the apparent simplicity of the construction, for certain $\mathcal{A}$, $\tilde{V}$ is not definable in an algebraically closed field; that is, $\tilde{V}(\mathcal{A})$ *is not classical* (Proposition 5.3.7).

Theorem B requires much more work, mainly the analysis of definable sets. This is because the theory of $\tilde{V}$, unlike the case of Zariski geometries coming from algebraic geometry, does not have quantifier elimination in the natural algebraic language. We hope that this technical analysis will be instrumental in practical applications to non-commutative geometry.

### 5.3.1 Algebras at roots of unity

The assumptions on $\mathcal{A}$ which allow us to carry out all the steps of the construction are listed here. There is a good chance that every known **quantum algebra at roots of unity** satisfies these assumptions or a modified version of these which still is sufficient for our construction. Note that there is no definition of quantum algebras at roots of unity, only a list of examples under the accepted common title. We give examples of a few such algebras satisfying our assumptions and invite the reader to check if the assumptions here cover all the cases of quantum algebras at roots of unity.

We fix until the end of the section a $K$-algebra $\mathcal{A}$ satisfying the following assumptions.

1. We assume that $K$ is an algebraically closed field and $\mathcal{A}$ is an associative unital affine $K$-algebra with generators $\mathbf{U}_1, \ldots, \mathbf{U}_d$ and defining relations with parameters in a finite $C \subset K$. We also assume that $\mathcal{A}$ is a finite dimensional module over its central sub-algebra $\mathcal{Z}$.

2. $\mathcal{Z}$ is a unital finitely generated commutative $K$-algebra without zero divisors, so Max $\mathcal{Z}$, the space of maximal ideals of $\mathcal{Z}$, can be identified with the $K$-points of an irreducible affine algebraic variety V over $C$.

3. There is a positive integer $N$ such that to every $m \in$ Max $\mathcal{Z}$ we can put in corresponds with $m$, an $\mathcal{A}$-module $M_m$ of dimension $N$ over $K$ with the property that the maximal ideal $m$ annihilates $M_m$.

   The isomorphism type of the module $M_m$ is determined uniformly by a solution to a system of polynomial equations $P^A$ in variables $t_{ijk} \in K$ and $m \in$ V such that for every $m \in$ V there exists $t = \{t_{ijk} : i \le d, \ j, k \le N\}$ satisfying $P^A(t, m) = 0$ and for each such $t$ there is a basis $e(1), \ldots, e(N)$ of the $K$-vector space on $M_m$ with

$$\bigwedge_{i \le d, \ j \le N} \mathbf{U}_i \, e(j) = \sum_{k=1}^{N} t_{ijk} e(k).$$

   We call any such basis $e(1), \ldots, e(N)$ **canonical**.

4. There is a finite group $\Gamma$ and a map $g : \text{V} \times \Gamma \to \text{GL}_N(K)$ such that, for each $\gamma \in \Gamma$, the map $g(\cdot, \gamma) : \text{V} \to \text{GL}_N(K)$ is rational $C$-definable (defined on an open subset of V) and, for any $m \in$ V, Dom$_m$, the domain of definition of the map $g(m, \cdot) : \Gamma \to \text{GL}_N(K)$, is a subgroup of $\Gamma$, $g(m, \cdot)$ is an injective homomorphism on its domain, and for any two canonical bases $e(1), \ldots, e(N)$ and $e'(1), \ldots, e'(N)$ of $M_m$ there is $\lambda \in K^*$ and $\gamma \in \text{Dom}_m$ such that

$$e'(i) = \lambda \sum_{1 \le j \le N} g_{ij}(m, \gamma) e(j), \quad i = 1, \ldots, N.$$

We denote

$$\Gamma_m := g(m, \text{Dom}_m).$$

**Remark 5.3.1.** The correspondence $m \mapsto M_m$ between points in V and the isomorphism types of modules is bijective by assumption 2. Indeed, for distinct $m_1, m_2 \in$ Max $\mathcal{Z}$, the modules $M_{m_1}$ and $M_{m_2}$ are not isomorphic, because otherwise the module will be annihilated by $\mathcal{Z}$.

### The associated structure

Recall that $\text{V}(\mathcal{A})$ or simply V stands for the $K$-points of the algebraic variety Max $\mathcal{Z}$. By assumption 1 in Section 5.3.1 this can be viewed as the set of $\mathcal{A}$-modules $M_m$, $m \in \mathcal{Z}$.

Consider the set $\tilde{V}$ as the disjoint union

$$\tilde{V} = \coprod_{m \in V} M_m.$$

For each $m \in V$, we also pick up arbitrarily a canonical basis $e = \{e(1), \ldots, e(N)\}$ in $M_m$ and all the other canonical bases conjugated to $e$ by $\Gamma_m$. We denote the set of bases for each $m \in V$ as

$$E_m := \Gamma_m e = \left\{ (e'(1), \ldots, e'(N)) : e'(i) = \sum_{1 \le j \le N} \gamma_{ij} e(j), \ \gamma \in \Gamma_m \right\}.$$

Consider, along with the sort $\tilde{V}$, also the field sort $K$, the sort $V$ identified with the corresponding affine sub-variety $V \subseteq K^k$, some $k$, and the projection map

$$\pi : x \mapsto m \qquad \text{if } x \in M_m \text{ from } \tilde{V} \text{ to } V.$$

We assume the *full language of* $\tilde{V}$ contains

1. the ternary relation $S(x, y, z)$ which holds iff there is $m \in V$ such that $x, y, z \in M_m$ and $x + y = z$ in the module;
2. the ternary relation $a \cdot x = y$ which for $a \in K$ and $x, y \in M_m$ is interpreted as the multiplication by the scalar $a$ in the module $M_m$;
3. the binary relations $\mathbf{U}_i x = y$, $(i = 1, \ldots, d)$ which for $x, y \in M_m$ are interpreted as the actions by the corresponding operators in the module $M_m$; and
4. the relations $E \subseteq V \times \tilde{V}^N$ with $E(m, e)$ interpreted as $e \in E_m$.

The *weak language* is the sublanguage of the full one which includes only 1–3.

Finally, denote $\tilde{V}$ the three-sorted structure $(\tilde{V}, V, K)$ described previously, with $V$ endowed with the usual Zariski language as the algebraic variety.

**Remark 5.3.2.**

1. Notice that the sorts $V$ and $K$ are bi-interpretable over $C$.
2. The map $g : V \times \Gamma \to \mathrm{GL}_N(K)$, being rational, is definable in the weak language of $\tilde{V}$.

Now we introduce the first-order theory $\mathrm{Th}(\mathcal{A}\text{-mod})$ describing $(\tilde{V}, V, K)$. It consists of axioms:

**Ax 1.** $K$ is an algebraically closed field of characteristic $p$ and $V$ is the Zariski structure on the $K$-points of the variety $\mathrm{Max}\, \mathcal{Z}$.

Ax 2. For each $m \in V$, the action of scalars of $K$ and operators $U_1, \ldots, U_d$ defines on $\pi^{-1}(m)$ the structure of an $\mathcal{A}$-module of dimension $N$.

Ax 3. Assumption 3 of Section 5.3.1 holds for the given $P^A$.

Ax 4. For the $g : V \times \Gamma \to GL_N(K)$ given by assumption 4 of Section 5.3.1, for any $e, e' \in E_m$ there exists $\gamma \in \Gamma$ such that

$$e'(i) = \sum_{1 \le j \le N} g_{ij}(m, \gamma)e(j), \quad i = 1, \ldots, N.$$

Moreover, $E_m$ is an orbit under the action of $\Gamma_m$.

**Remark 5.3.3.** Note that if $M_m$ is irreducible, then associated to a particular collection of coefficients $t_{kij}$ there is a unique (up to scalar multiplication) canonical base for $M_m$ (as in Exercise 2.1.3). It follows that the only possible automorphisms of $\tilde{V}$ which fix all of $F$ are induced by multiplication by scalars in each module. (The scalars do not have to be the same for each fibre and typically are not.) So the 'projective' bundle $\coprod_{m \in V}(M_m/\text{scalars})$ is internal to the field $K$, but the original $\tilde{V}$ is not.

## 5.3.2 Examples

We assume that $\epsilon \in K$ is a primitive root of 1 of order $\ell$ and $\ell$ is not divisible by the characteristic of $K$.

0. Let $\mathcal{A}$ be a commutative unital affine $K$-algebra. We let $\mathcal{Z} = \mathcal{A}$, and so $V = \text{Max } \mathcal{Z} = \text{Max } \mathcal{A}$ is the corresponding affine variety. Ideals of $m \in \text{Max } \mathcal{A}$ annihilate irreducible one-dimensional (over $K$) $\mathcal{A}$-modules $M_m$, and this gives us a trivial line bundle $\{M_m : m \in V\}$. Triviality means that the bundle is definable in $K$ in the sense of algebraic geometry and we have a section that is a rational map

$$s : V \to \coprod_{m \in V} M_m.$$

The $s(m)$ can be considered a canonical basis of $M_m$ for every $m \in V$. $\Gamma_m$ is the unit group for all $m$.

In other words, for a commutative algebra, $\tilde{V}$ is just the affine variety Max $\mathcal{A}$ equipped with a trivial linear bundle.

1. Let $\mathcal{A}$ be generated by $U, V, U^{-1}, V^{-1}$ satisfying the relations

$$UU^{-1} = 1 = VV^{-1}, \qquad UV = \epsilon VU.$$

We denote this algebra $T_\epsilon^2$ [equivalent to $\mathcal{O}_\epsilon((K^\times)^2)$] in Brown and Goodearl's notations (2002).

The centre $Z = \mathcal{Z}$ of $T_\epsilon^2$ is the sub-algebra generated by $\mathbf{U}^\ell, \mathbf{U}^{-\ell}, \mathbf{V}^\ell, \mathbf{V}^{-\ell}$. The variety Max $Z$ is isomorphic to the two-dimensional torus $K^* \times K^*$.

Any irreducible $T_\epsilon^2$-module $M$ is a $K$-vector space of dimension $N = \ell$. It has a basis $\{e_0, \ldots, e_{\ell-1}\}$ of the space consisting of $\mathbf{U}$-eigenvectors and satisfying, for an eigenvalue $\mu$ of $\mathbf{U}$ and an eigenvalue $v$ of $\mathbf{V}$,

$$\mathbf{U}e_i = \mu\epsilon^i e_i$$
$$\mathbf{V}e_i = \begin{cases} ve_{i+1}, & i < \ell - 1, \\ ve_0, & i = \ell - 1. \end{cases}$$

We also have a basis of $\mathbf{V}$-eigenvectors $\{g_0, \ldots, g_{\ell-1}\}$ satisfying

$$g_i = e_0 + \epsilon^i e_1 + \cdots + \epsilon^{i(\ell-1)} e_{\ell-1},$$

and so

$$\mathbf{V}g_i = v\epsilon^i g_i$$
$$\mathbf{U}g_i = \begin{cases} \mu g_{i+1}, & i < \ell - 1, \\ \mu g_0, & i = \ell - 1. \end{cases}$$

For $\mu^\ell = a \in K^*$ and $v^\ell = b \in K^*$, $(\mathbf{U}^\ell - a), (\mathbf{V}^\ell - b)$ are generators of $\mathrm{Ann}(M)$. The module is determined uniquely once the values of $a$ and $b$ are given. So, V is isomorphic to the two-dimensional torus $K^* \times K^*$.

The coefficients $t_{ijk}$ in this example are determined by $\mu$ and $v$, which satisfy the polynomial equations $\mu^\ell = a$, $v^\ell = b$.

$\Gamma_m = \Gamma$ is the fixed nilpotent group of order $\ell^3$ generated by the matrices

$$\begin{pmatrix} 0\,1\,0\ldots0 \\ 0\,0\,1\ldots0 \\ \cdots\ \cdots \\ 1\,0\ \ldots0 \end{pmatrix} \quad \text{and} \quad \begin{pmatrix} 1\,0\,0\ldots0 \\ 0\,\epsilon\,0\ldots0 \\ \cdots\ \cdots \\ 0\,0\ldots\epsilon^{\ell-1} \end{pmatrix}$$

2. Similarly, the $d$-dimensional quantum torus $T_{\epsilon,\theta}^d$ generated by $\mathbf{U}_1, \ldots, \mathbf{U}_d$, $\mathbf{U}_1^{-1}, \ldots, \mathbf{U}_d^{-1}$ satisfying

$$\mathbf{U}_i \mathbf{U}_i^{-1} = 1, \qquad \mathbf{U}_i \mathbf{U}_j = \epsilon^{\theta_{ij}} \mathbf{U}_j \mathbf{U}_i, \qquad 1 \le i, j \le d,$$

where $\theta$ is an antisymmetric integer matrix, g.c.d.$\{\theta_{ij} : 1 \le j \le d\}) = 1$ for some $i \le d$.

There is a simple description of the bundle of irreducible modules, all of which are of the same dimension $N = \ell$.

$T_{\epsilon,\theta}^d$ satisfies all the assumptions.

3. $\mathcal{A} = U_\epsilon(\mathfrak{sl}_2)$, the quantum universal enveloping algebra of $SL_2(K)$. It is given by generators $K, K^{-1}, E, F$, satisfying the defining relations

$$KK^{-1} = 1, \qquad KEK^{-1} = \epsilon^2 E, \qquad KFK^{-1} = \epsilon^{-2} F,$$

$$EF - FE = \frac{K - K^{-1}}{\epsilon - \epsilon^{-1}}.$$

The centre $Z$ of $U_\epsilon(\mathfrak{sl}_2)$ is generated by $K^\ell, E^\ell, F^\ell$, and the element

$$C = FE + \frac{K\epsilon + K^{-1}\epsilon^{-1}}{(\epsilon - \epsilon^{-1})^2}.$$

We use Brown and Goodearl's notation (2002), Chapter III.2, to describe $\tilde{V}$. We assume $\ell \geq 3$ odd.

Let $\mathcal{Z} = Z$, and so $V = \text{Max } Z$ is an algebraic extension of degree $\ell$ of the commutative affine algebra $K^\ell, K^{-\ell}, E^\ell, F^\ell$.

To every point, $m = (a, b, c, d) \in V$ corresponds the unique, up-to-isomorphism module with a canonical basis $e_0, \ldots, e_{\ell-1}$, satisfying

$$Ke_i = \mu\epsilon^{-2i} e_i,$$

$$Fe_i = \begin{cases} e_{i+1}, & i < \ell - 1, \\ be_0, & i = \ell - 1, \end{cases}$$

$$Ee_i = \begin{cases} \rho e_{\ell-1}, & i = 0, \\ (\rho b + \frac{(\epsilon^i - \epsilon^{-i})(\mu\epsilon^{1-i} - \mu^{-1}\epsilon^{i-1})}{(\epsilon - \epsilon^{-1})^2})e_{i-1}, & i > 0, \end{cases}$$

where $\mu, \rho$ satisfy the polynomial equations

$$\mu^\ell = a, \qquad \rho b + \frac{\mu\epsilon + \mu^{-1}\epsilon^{-1}}{(\epsilon - \epsilon^{-1})^2} = d \qquad (5.16)$$

and

$$\rho \prod_{i=1}^{\ell-1} \left( \rho b + \frac{(\epsilon^i - \epsilon^{-i})(\mu\epsilon^{1-i} - \mu^{-1}\epsilon^{i-1})}{(\epsilon - \epsilon^{-1})^2} \right) = c. \qquad (5.17)$$

We may characterise $V$ as

$$V = \{(a, b, c, d) \in K^4 : \exists \, \rho, \mu \, (5.16) \text{ and } (5.17) \text{ hold}\}.$$

In fact, the map $(a, b, c, d) \mapsto (a, b, c)$ is a cover of the affine variety $A^3 \cap \{a \neq 0\}$ of order $\ell$.

In almost all points of $V$, except for the points of the form $(1, 0, 0, d_+)$ and $(-1, 0, 0, d_-)$, the module is irreducible. In the exceptional cases, for each $i \in \{0, \ldots, \ell - 1\}$, we have exactly one $\ell$-dimensional module [denoted $\mathcal{Z}(\epsilon^i)$ or $\mathcal{Z}(-\epsilon^i)$ in Brown and Goodearl (2002), depending on the sign] which satisfies the description with $\mu = \epsilon^i$ or $-\epsilon^i$. The Casimir invariant is

$$d_+ = \frac{\epsilon^{i+1} + \epsilon^{-i-1}}{(\epsilon - \epsilon^{-1})^2} \quad \text{or} \quad d_- = -\frac{\epsilon^{i+1} + \epsilon^{-i-1}}{(\epsilon - \epsilon^{-1})^2}$$

and the module, for $i < \ell - 1$, has the unique proper irreducible sub-module of dimension $\ell - i - 1$ spanned by $e(i + 1), \ldots, e(\ell - 1)$. For $i = \ell - 1$, the module is irreducible. According to Brown and Goodearl (2002), in Chapter III.2, all the irreducible modules of $\mathcal{A}$ have been listed, either as $M_m$ or as sub-modules of $M_m$ for the exceptional $m \in V$.

To describe $\Gamma_m$, consider two canonical bases $e$ and $e'$ in $M_m$. If $e'$ is not of the form $\lambda e$, then necessarily $e'_0 = \lambda e_k$, for some $k \leq \ell - 1$, $b \neq 0$, and

$$e'_i = \begin{cases} \lambda e_{i+k}, & 0 \leq i < \ell - k, \\ \lambda b e_{i+k}, & \ell - 1 \geq i \geq \ell - k. \end{cases}$$

If we put $\lambda = \lambda_k = \nu^{-k}$, for $\nu^\ell = b$, we get a finite-order transformation. We can take $\Gamma_{(a,b,c,d)}$, for $b \neq 0$, to be the Abelian group of order $\ell^2$ generated by the matrices

$$\begin{pmatrix} 0 & \nu^{-1} & 0 & \ldots & 0 \\ 0 & 0 & \nu^{-1} & \ldots & 0 \\ \ldots & \ldots & & & \nu^{-1} \\ \nu^{\ell-1} & 0 & & \ldots & 0 \end{pmatrix} \quad \text{and} \quad \begin{pmatrix} \epsilon & 0 & 0 & \ldots & 0 \\ 0 & \epsilon & 0 & \ldots & 0 \\ \ldots & & \ldots & & \\ 0 & 0 & & \ldots & \epsilon \end{pmatrix}$$

where $\nu$ is defined by

$$\nu^\ell = b.$$

When $b = 0$, the group $\Gamma_{(a,0,c,d)}$ is just the cyclic group generated by the scalar matrix with $\epsilon$ on the diagonal.

*The isomorphism type of the module depends on $\langle a, b, c, d \rangle$ only.* This basis satisfies assumptions 1–4.

$U_\epsilon(\mathfrak{sl}_2)$ is one of the simplest examples of a *quantum group*. Quantum groups, as all bi-algebras, have the following crucial property: *the tensor product $M_1 \otimes M_2$ of any two $\mathcal{A}$-modules is well defined and is an $\mathcal{A}$-module.* So, the tensor product of two modules in $\tilde{V}$ produces a $U_\epsilon(\mathfrak{sl}_2)$-module of dimension $\ell^2$, definable in the structure, and which 'contains' finitely many modules in $\tilde{V}$. This defines a multivalued operation on $V$ (or on the open subset of $V$, in the second case).

More examples and the most general known cases $U_\epsilon(\mathfrak{g})$, for $\mathfrak{g}$ a semi-simple complex Lie algebra, and $\mathcal{O}_\epsilon(G)$, the quantised group $G$, for $G$, a connected simply connected semi-simple complex Lie group, are shown to have properties 1 and 2 for the central algebra $\mathcal{Z}$ generated by the corresponding $U_i^\ell, i = 1, \ldots, d$.

The rest of the assumptions are harder to check. We leave this open.

4. $\mathcal{A} = \mathcal{O}_\epsilon(K^2)$; Manin's quantum plane is given by generators $\mathbf{U}$ and $\mathbf{V}$ and defining relations $\mathbf{UV} = \epsilon \mathbf{VU}$. The centre $\mathcal{Z}$ is again generated by $\mathbf{U}^\ell$, and $\mathbf{V}^\ell$, and the maximal ideals of $Z$ in this case are of the form $\langle (\mathbf{U}^\ell - a), (\mathbf{V}^\ell - b) \rangle$ with $\langle a, b \rangle \in K^2$.

This example, though very easy to understand algebraically, does not quite fit into our construction. Namely, assumption 3 is satisfied only in generic points of $V = \operatorname{Max} \mathcal{Z}$, but the main statement still holds true for this case as well. We just have to construct $\tilde{V}$ by glueing two Zariski spaces, each corresponding to a localisation of the algebra $\mathcal{A}$.

To each maximal ideal with $a \neq 0$, we put in correspondence the module of dimension $\ell$ given in a basis $e_0, \ldots, e_{\ell-1}$ by

$$\mathbf{U}e_i = \mu \epsilon^i e_i$$
$$\mathbf{V}e_i = \begin{cases} e_{i+1}, & i < \ell - 1, \\ be_0, & i = \ell - 1 \end{cases}$$

for $\mu$ satisfying $\mu^\ell = a$.

To each maximal ideal with $b \neq 0$, we put in correspondence the module of dimension $\ell$ given in a basis $g_0, \ldots, g_{\ell-1}$ by

$$\mathbf{V}g_i = \nu \epsilon^i g_i$$
$$\mathbf{U}g_i = \begin{cases} g_{i+1}, & i < \ell - 1, \\ ae_0, & i = \ell - 1 \end{cases}$$

for $\nu$ satisfying $\nu^\ell = b$.

When both $a \neq 0$ and $b \neq 0$, we identify the two representations of the same module by choosing $g$ (given $e$ and $\nu$) so that

$$g_i = e_0 + \nu^{-1} \epsilon^i e_1 + \cdots + \nu^{-k} \epsilon^{ik} e_k + \cdots + \nu^{-(\ell-1)} \epsilon^{i(\ell-1)} e_{\ell-1}.$$

This induces a definable isomorphism between modules and defines a glueing between $\tilde{V}_{a \neq 0}$ and $\tilde{V}_{b \neq 0}$. In fact, $\tilde{V}_{a \neq 0}$ corresponds to the algebra given by three generators $\mathbf{U}, \mathbf{U}^{-1}$, and $\mathbf{V}$ with relations $\mathbf{UV} = \epsilon \mathbf{VU}$ and $\mathbf{UU}^{-1} = 1$, a localisation of $\mathcal{O}_\epsilon(K^2)$, and $\tilde{V}_{b \neq 0}$ corresponds to the localisation by $\mathbf{V}^{-1}$.

## Categoricity

**Lemma 5.3.4.**

(i) *Let $\tilde{V}_1$ and $\tilde{V}_2$ be two structures in the weak language satisfying assumptions 1–3 in Section 5.3.1 with the same $P^A$ over the same algebraically closed field $K$. Then the natural isomorphism $i : V_1 \cup K \to V_2 \cup K$ over $C$ can be lifted to an isomorphism*

$$i : \tilde{V}_1 \to \tilde{V}_2.$$

(ii) *Let $\tilde{V}_1$ and $\tilde{V}_2$ be two structures in the full language satisfying assumptions 1–4 in Section 5.3.1 with the same $P^A$ over the same algebraically closed field $K$. Then the natural isomorphism $i : V_1 \cup K \to V_2 \cup K$ over $C$ can be lifted to an isomorphism*

$$i : \tilde{V}_1 \to \tilde{V}_2.$$

*Proof.* We may assume that $i$ is the identity on V and on the sort $K$.

The assumption 1 and the description 1 in Section 5.3.1 imply that in both structures, $\pi^{-1}(m)$, for $m \in V$, has the structure of a module. Denote these $\pi_1^{-1}(m)$ and $\pi_2^{-1}(m)$ in the first and second structure, respectively.

For each $m \in V$, the modules $\pi_1^{-1}(m)$ and $\pi_2^{-1}(m)$ are isomorphic.

Indeed, by using assumption 3 in Section 5.3.1, choose $t_{ijk}$ satisfying $P^A$ for $m$ and find bases $e$ in $\pi_1^{-1}(m)$ and $e'$ in $\pi_2^{-1}(m)$ with the $\mathbf{U}_i$'s represented by the matrices $\{t_{ijk} : k, j = 1, \ldots, N\}$ in both modules. It follows that the map

$$i_m : \sum z_j e(j) \mapsto \sum z_j e'(j), \qquad z_1, \ldots, z_N \in K$$

is an isomorphism of the $\mathcal{A}$-modules

$$i_m : \pi_1^{-1}(m) \to \pi_2^{-1}(m).$$

Hence, the union

$$\mathbf{i} = \bigcup_{m \in V} i_m, \qquad \mathbf{i} : \tilde{V}_1 \to \tilde{V}_2,$$

is an isomorphism. This proves (i).

To prove (ii), choose, using assumption 4 in Section 5.3.1, $e$ and $e'$ in $E_m$ in $\pi_1^{-1}(m)$ and $\pi_2^{-1}(m)$. Then the map $i_m$ by the same assumption also preserves $E_m$, and so $\mathbf{i}$ is an isomorphism in the full language. $\qquad\square$

As an immediate corollary we get the following.

**Theorem 5.3.5.** Th($\mathcal{A}$-mod) *is categorical in uncountable cardinals both in the full and the weak languages.*

**Remark 5.3.6.** The previous lemma is a special case of Lemma 5.3.9 in the next sub-section.

We now prove that despite the simplicity of the construction and the proof of categoricity, the structures obtained from algebras $\mathcal{A}$ in our list of examples are non-classical.

Assume for simplicity that char $K = 0$. The statements in this sub-section are in their strongest form when we choose the weak language for the structures.

**Proposition 5.3.7.** $\tilde{V}(T_\epsilon^n)$ *is not definable in an algebraically closed field, for* $n \geq 2$.

*Proof.* We write $\mathcal{A}$ for $T_\epsilon^2$. We consider the structure in the weak language.

Suppose towards the contradiction that $\tilde{V}(\mathcal{A})$ is definable in some $K'$. Then $K$ is also definable in this algebraically closed field. But, as is well known, the only infinite field definable in an algebraically closed field is the field itself. So, $K' = K$, and we have to assume that $\tilde{V}$ is definable in $K$.

Given $\mathbf{W} \in \mathcal{A}$, $v \in \tilde{V}$, $x \in K$, and $m \in V$, by $\mathrm{Eig}(\mathbf{W}; v, x, m)$ denote the statement $v$ *is an eigenvector of* $\mathbf{W}$ *in* $\pi^{-1}(m)$ *(or simply in* $M_m$*) with the eigenvalue* $x$.

For any given $\mathbf{W}$, the ternary relation $\mathrm{Eig}(\mathbf{W}; v, x, m)$ is definable in $\tilde{V}$ by 5.3.1.

Let $m \in V$ be such that $\mu$ is a $\mathbf{U}$-eigenvalue and $\nu$ is a $\mathbf{V}$-eigenvalue in the module $M_m$. Then $\langle \mu^\ell, \nu^\ell \rangle$ determines the isomorphism type of $M_m$ (see Section 5.3.2); in fact, $m = \langle \mu^\ell, \nu^\ell \rangle$.

Consider the definable set

$$\mathrm{Eig}(\mathbf{U}) = \{v \in \tilde{V} : \exists \mu, m\,\mathrm{Eig}(\mathbf{U}; v, \mu, m)\}.$$

By our assumption and elimination of imaginaries in ACF, this is in a definable bijection with an algebraic subset $S$ of $K^n$, some $n$, defined over some finite $C'$. We may assume that $C' = C$. Moreover, the relations and functions induced from $\tilde{V}$ on $\mathrm{Eig}(\mathbf{U})$ are algebraic relations definable in $K$ over $C$.

Consider $\mu$ and $\nu$ as variables running in $K$ and let $\tilde{K} = K\{\mu, \nu\}$ be the field of Puiseux series in variables $\mu, \nu$. Because $S(\tilde{K})$ as a structure is an elementary extension of $\mathrm{Eig}(\mathbf{U})$, there is a tuple, say $e_\mu$, in $S(\tilde{K})$ which is a $\mathbf{U}$-eigenvector with the eigenvalue $\mu$.

By definition, the co-ordinates of $e_\mu$ are Laurent series in the variables $\mu^{\frac{1}{k}}$ and $\nu^{\frac{1}{k}}$, for some positive integer $k$. Let $K$ be the sub-field of $\tilde{K}$ consisting of all Laurent series in variables $\mu^{\frac{1}{k}}, \nu^{\frac{1}{k}}$, for the $k$. Fix $\delta \in K$ such that

$$\delta^k = \epsilon.$$

The maps

$$\xi : t(\mu^{\frac{1}{k}}, v^{\frac{1}{k}}) \mapsto t(\delta\mu^{\frac{1}{k}}, v^{\frac{1}{k}}) \qquad \text{and} \qquad \zeta : t(\mu^{\frac{1}{k}}, v^{\frac{1}{k}}) \mapsto t(\mu^{\frac{1}{k}}, \delta v^{\frac{1}{k}}),$$

for $t(\mu^{\frac{1}{k}}, v^{\frac{1}{k}})$ Laurent series in the corresponding variables, obviously are automorphisms of $K$ over $K$. In particular, $\xi$ maps $\mu$ to $\epsilon\mu$ and leaves $v$ fixed, and $\zeta$ maps $v$ to $\epsilon v$ and leaves $\mu$ fixed. Also note that the two automorphisms commute and both are of order $\ell k$.

Because $\mathbf{U}$ is $K$-definable, $\xi^m(e_\mu)$ is a $\mathbf{U}$-eigenvector with the eigenvalue $\epsilon^m\mu$, for any integer $m$.

By the properties of $\mathcal{A}$-modules, $\mathbf{V}e_\mu$ is a $\mathbf{U}$-eigenvector with the eigenvalue $\epsilon\mu$, so there is $\alpha \in \tilde{K}$

$$\mathbf{V}e_\mu = \alpha\xi(e_\mu). \tag{5.18}$$

But $\alpha$ is definable in terms of $e_\mu$, $\xi(e_\mu)$, and $C$, so by elimination of quantifiers $\alpha$ is a rational function of the coordinates of the elements; hence, $\alpha \in K$.

Because $\mathbf{V}$ is definable over $K$, we have for every automorphism $\gamma$ of $K$,

$$\gamma(\mathbf{V}e) = \mathbf{V}\gamma(e).$$

So, Equation (5.18) implies

$$\mathbf{V}\xi^i e_\mu = \xi^i(\alpha)\xi^{i+1}(e_\mu), \qquad i = 0, 1, 2, \dots$$

and, because

$$\mathbf{V}^{k\ell}e_\mu = v^{k\ell}e_\mu,$$

by applying $\mathbf{V}$ to both sides of Equation (5.18) $k\ell - 1$ times we get

$$\prod_{i=0}^{k\ell-1} \xi^i(\alpha) = v^{k\ell}. \tag{5.19}$$

Now remember that

$$\alpha = a_0(v^{\frac{1}{k}}) \cdot \mu^{\frac{d}{k}} \cdot (1 + a_1(v^{\frac{1}{k}})\mu^{\frac{1}{k}} + a_2(v^{\frac{1}{k}})\mu^{\frac{2}{k}} + \cdots)$$

where $a_0(v^{\frac{1}{k}}), a_1(v^{\frac{1}{k}}), a_2(v^{\frac{1}{k}}), \dots$ are Laurent series in $v^{\frac{1}{k}}$ and $d$ an integer. By substituting this into Equation (5.19), we get

$$v^{k\ell} = a_0(v^{\frac{1}{k}})^{k\ell}\delta^{\frac{k\ell(k\ell-1)}{2}}\mu^{d\ell} \cdot (1 + a_1'(v^{\frac{1}{k}})\mu^{\frac{1}{k}} + a_2'(v^{\frac{1}{k}})\mu^{\frac{2}{k}} + \cdots).$$

It follows that $d = 0$ and $a_0(v^{\frac{1}{k}}) = a_0 \cdot v$, for some constant $a_0 \in K$. That is,

$$\alpha = a_0 \cdot v \cdot (1 + a_1(v^{\frac{1}{k}})\mu^{\frac{1}{k}} + a_2(v^{\frac{1}{k}})\mu^{\frac{2}{k}} + \cdots). \tag{5.20}$$

Now we use the fact that $\zeta(e_\mu)$ is a **U** eigenvector with the same eigenvalue $\mu$, so by the same argument as before there is $\beta \in K$ such that

$$\zeta(e_\mu) = \beta e_\mu. \tag{5.21}$$

So,

$$\zeta^{i+1}(e_\mu) = \zeta^i(\beta)\zeta^i(e_\mu)$$

and taking into account that $\zeta^{k\ell} = 1$, we get

$$\prod_{i=0}^{k\ell-1} \zeta^i(\beta) = 1.$$

Again we analyse $\beta$ as a Laurent series and represent it in the form

$$\beta = b_0(\mu^{\frac{1}{k}}) \cdot v^{\frac{d}{k}} \cdot (1 + b_1(\mu^{\frac{1}{k}})v^{\frac{1}{k}} + b_2(\mu^{\frac{1}{k}})v^{\frac{2}{k}} + \cdots)$$

where $b_0(\mu^{\frac{1}{k}}), b_1(\mu^{\frac{1}{k}}), b_2(\mu^{\frac{1}{k}}), \ldots$ are Laurent series of $\mu^{\frac{1}{k}}$ and $d$ is an integer.

By an argument similar to the previous one, using Equation (5.22) we get

$$\beta = b_0 \cdot (1 + b_1(\mu^{\frac{1}{k}})v^{\frac{1}{k}} + b_2(\mu^{\frac{1}{k}})v^{\frac{2}{k}} + \cdots) \tag{5.22}$$

for some $b_0 \in K$.

Finally, we use the fact that $\xi$ and $\zeta$ commute. By applying $\zeta$ to Equation (5.18), we get

$$\mathbf{V}\zeta(e_\mu) = \zeta(\alpha)\zeta\xi(e_\mu) = \zeta(\alpha)\xi\zeta(e_\mu) = \xi(\beta)\zeta(\alpha)\xi(e_\mu).$$

On the other hand,

$$\mathbf{V}\zeta(e_\mu) = \beta\mathbf{V}e_\mu = \beta\alpha\xi(e_\mu).$$

That is,

$$\frac{\alpha}{\zeta(\alpha)} = \frac{\xi(\beta)}{\beta}.$$

By substituting Equations (5.20) and (5.22) and dividing on both sides, we get the equality

$$\epsilon^{-1}(1 + a_1'(v^{\frac{1}{k}})\mu^{\frac{1}{k}} + a_2'(v^{\frac{1}{k}})\mu^{\frac{2}{k}} + \cdots) = 1 + b_1'(\mu^{\frac{1}{k}})v^{\frac{1}{k}} + b_2'(\mu^{\frac{1}{k}})v^{\frac{2}{k}} + \cdots$$

By comparing the constant terms on both sides, we get the contradiction. This proves the proposition in the case $n = 2$.

To end the proof, we just notice that the structure $\tilde{\mathbf{V}}(T_\epsilon^2)$ is definable in any of the other $\tilde{\mathbf{V}}(T_\epsilon^n)$, maybe with a different root of unity. This follows from the fact that the $\mathcal{A}$-modules in all cases have similar description. $\qquad\square$

**Corollary 5.3.8.** *The structure* $V(U_\epsilon(\mathfrak{sl}_2))$ *(Example 3 in Section 5.3.2) is not definable in an algebraically closed field.*

Indeed, consider

$$V_0 = \{(a, b, c, d) \in V : b \neq 0, \ c = 0\} \qquad \text{and} \qquad \tilde{V}_0 = \pi^{-1}(V_0)$$

with the relations induced from $\tilde{V}$.

Set $U := K$, $V = F$, and consider the reduct of the structure $\tilde{V}_0$ which ignores the operators $E$ and $C$. This structure is isomorphic to $\tilde{V}(T_{\epsilon^2}^2)$ and is definable in $V(U_\epsilon(\mathfrak{sl}_2))$, so the latter is not definable in an algebraically closed field. $\qquad\square$

### 5.3.3 Definable sets and Zariski properties

#### Canonical formulas

Given variables $v_{1,1}, \ldots, v_{1,r_1}, \ldots, v_{s,1}, \ldots, v_{s,r_s}$ of the sort $\tilde{V}$, $m_1, \ldots, m_s$ of the sort V, and variables $x = \{x_1, \ldots, x_p\}$ of the sort $K$, denote by $A_0(e, m, t)$ the formula

$$\bigwedge_{i \leq s, \ j \leq N} E(e_i, m_i) \ \& \ P^A(\{t_{ikn\ell}\}_{k \leq d, \ \ell, n \leq N}; m_i) = 0 \ \&$$

$$\bigwedge_{k \leq d, j \leq N, i \leq s} U_k e_i(j) = \sum_{\ell \leq N} t_{ikj\ell} e_i(\ell).$$

Denote by $A(e, m, t, z, v)$ the formula

$$A_0(e, m, t) \ \& \bigwedge_{i \leq s; \ j \leq r_i} v_{ij} = \sum_{\ell \leq N} z_{ij\ell} e_i(\ell).$$

The formula of the form

$$\exists \, e_1, \ldots, e_s \exists \, m_1, \ldots, m_s$$

$$\exists \, \{t_{ikjl} : k \leq d, \ i \leq s, \ j, \ell \leq N\} \subseteq K$$

$$\exists \, \{z_{ijl} : i \leq s, \ j \leq r_i, \ \ell \leq N\} \subseteq K :$$

$$A(e, m, t, z, v) \ \& \ R(m, t, x, z),$$

where $R$ is a Boolean combination of Zariski closed predicates in the algebraic variety $V^s \times K^q$ over $C$, $q = |t| + |x| + |z|$ (constructible predicate over $C$) is called **a core $\exists$-formula with kernel** $R(m, t, x, z)$ **over** $C$. The enumeration of variables $v_{ij}$ is referred to as the **partitioning enumeration**.

We also refer to this formula as $\exists e R$.

**Comments**

(i) A core formula is determined by its kernel once the partition of variables (by enumeration) is fixed. The partition sets that $\pi(e_i(j)) = \pi(e_i(k))$, for every $i, j, k$, and fixes the components of the subformula $A(e, m, t, z, v)$.

(ii) The relation $A_0(e, m, t)$ defines the functions

$$e \mapsto (m, t),$$

that is, given a canonical basis $\{e_i(1), \ldots, e_i(N)\}$ in $M_{m_i}$ we can uniquely determine $m_i$ and $t_{ikj\ell}$.

For the same reason, $A(e, m, t, z, v)$ defines the functions

$$(e, v) \mapsto (m, t, z).$$

**Lemma 5.3.9.** *Let*

$$a = \langle a_{1,1}, \ldots, a_{1,r_1}, \ldots, a_{s,1}, \ldots, a_{s,r_s} \rangle \in \tilde{V} \times \ldots \times \tilde{V}, \ b = \langle b_1, \ldots, b_n \rangle \in K^n.$$

*The complete type* $\mathrm{tp}(a, b)$ *of the tuple over C is determined by its subtype* $\mathrm{ctp}(a, b)$ *over C, consisting of core ∃-formulas.*

*Proof.* We are going to prove that, given $a', b'$ satisfying the same core type $\mathrm{ctp}(a, b)$, there is an automorphism of any $\aleph_0$-saturated model, $\alpha : (a, b) \mapsto (a', b')$.

We assume that the enumeration of variables has been arranged so that $\pi(a_{ij}) = \pi(a_{kn})$ iff $i = k$. Denote $m_i = \pi(a_{ij})$.

Let $e_i$ be bases of modules $\pi^{-1}(m_i)$, $i = 1, \ldots, s$, $j = 1, \ldots, N$, such that $\models A_0(e, m, t)$ for some $t = \{t_{ikj\ell}\}$ (see the notation in Remark 5.3.3 and assumption 3 in Section 5.3.1), in particular $e_i \in E_{m_i}$. By the assumption that the correspondent systems span $M_{m_i}$, there exist $c_{ij\ell}$ such that

$$\bigwedge_{i \leq s; \ j \leq r_i} a_{ij} = \sum_{\ell \leq N} c_{ij\ell} e_i(\ell),$$

and let $p = \{P_i : i \in \mathbb{N}\}$ be the complete algebraic type of $(m, t, b, c)$.

The type $\mathrm{ctp}(a, b)$ contains core formulas with kernels $P_i$, $i = 1, 2, \ldots$ By assumptions and saturatedness, we can find $e'$ $m'$, $t'$, and $c'$ satisfying the corresponding relations for $(a', b')$. In particular, the algebraic types of $(m, t, b, c)$ and $(m', t', b', c')$ over $C$ coincide and $e_i' \in E_{m_i'}$. It follows that there is an automorphism $\alpha : K \to K$ over $C$ such that $\alpha : (m, t, b, c) \mapsto (m', t', b', c')$.

Extend $\alpha$ to $\pi^{-1}(m_1) \cup \cdots \cup \pi^{-1}(m_s)$ by setting

$$\alpha\left(\sum_j z_j e_i(j)\right) = \sum_j \alpha(z_j) e_i'(j) \tag{5.23}$$

for any $z_1, \ldots, z_N \in K$ and $i \in \{1, \ldots, s\}$. In particular $\alpha(a_{ij}) = a_{ij}'$ and, because $\alpha(\Gamma_{m_i}) = \Gamma_{m_i'}$, also $\alpha(E_{m_i}) = E_{m_i'}$.

Now, for each $m \in V \setminus \{m_1, \ldots, m_s\}$ we construct the extension of $\alpha$, $\alpha_m^+ : \pi^{-1}(m) \rightarrow \pi^{-1}(m')$, for $m' = \alpha(m)$, as in Lemma 5.3.4. Use assumption 3 in Section 5.3.1 to choose $t_{ijk}$ satisfying $P^A$ for $m$, and find bases $e \in E_m$ and $e' \in E_{m'}$ with the $U_i$'s represented by the matrices $\{t_{ijk} : k, j = 1, \ldots, N\}$ in $\pi^{-1}(m)$ and by $\{\alpha(t_{ijk}) : k, j = 1, \ldots, N\}$ in $\pi^{-1}(m')$. It follows that the map

$$\alpha_m^+ : \sum z_j e(j) \mapsto \sum \alpha(z_j) e'(j), \quad z_1, \ldots, z_N \in K$$

is an isomorphism of the $\mathcal{A}$-modules

$$\alpha_m^+ : \pi^{-1}(m) \rightarrow \pi^{-1}(m').$$

Hence, the union

$$\alpha^+ = \bigcup_{m \in V} \alpha_m^+$$

is an automorphism of $\tilde{V}$. $\qquad\square$

By the compactness theorem, we immediately get the following from the lemma.

**Corollary 5.3.10.** *Every formula in $\tilde{V}$ with parameters in $C \subseteq K$ is equivalent to the disjunction of a finite collection of core formulas.*

**Theorem 5.3.11.** *For any algebra $\mathcal{A}$ satisfying assumptions 1–4 in Section 5.3.1 the structure $\tilde{V}$ is a Zariski geometry, satisfying the pre-smoothness condition provided the affine algebraic variety V is smooth.*

*Proof.* Take sets defined by positive core formulas to be Zariski closed. Analysis of these allows one to check the axioms of a Zariski structure. See Zilber (2008a) for the detailed proof. $\qquad\square$

# 6

# Analytic Zariski geometries

The notion of an analytic Zariski structure was introduced in [**PZ**] by the author and N. Peatfield in a form slightly different from the one presented here. Analytic Zariski generalises the previously known notion of a Zariski structure of Chapters 3–5 mainly by dropping the requirement of Noetherianity and weakening the assumptions on the projections of closed sets. However, this is not just a technical generalisation. It opens the doors for two completely new classes of examples:

(i) structures which are constructed in terms of complex analytic functions and relations, and

(ii) 'new stable structures' introduced by the Hrushovski construction (see Section B.2.2) which in many cases exhibit properties similar to those of class (i).

It is also an attempt to treat the two classes of structures in a uniform way, revealing a common broad idea of what mathematicians mean by *analytic*. Indeed, the word *analytic* is used to describe different things in the complex and in the real context, as well as in the p-adic setting. More subtle but similar phenomena are encountered in the context of non-commutative and quantum geometry. We believe that the model-theoretic analysis undertaken in this chapter and in several related papers is a step in this direction.

## 6.1 Definition and basic properties

We introduce analytic Zariski structures as (non-Noetherian) topological structures with good dimension notion for all definable subsets, that is (DP), (FC), and (AF) hold in the same form as in Section 3.1.2 but for a wider family of sets. We change the semi-projectivity condition (SP) to a more general form consistent with its previous use. We also generalise (DU) and (EU) and add an important assumption (AS), the analytic stratification of closed sets.

The logician may notice that the logic formalism here shifts from the first-order context to that of infinitary languages, perhaps even to *abstract elementary classes,* although we do not elaborate on this.

### 6.1.1 Closed and projective sets

We assume our structure M to be a topological structure (of Section 2.1). Further on, we assume that M has a good dimension notion.

To any non-empty projective $S$, a non-negative integer called the **dimension of** $S$, dim $S$, is attached.

We assume (DP) and (SI) and strengthen, formally, (DU) to (CU):

(CU) **Countable unions:** If $S = \bigcup_{i \in \mathbb{N}} S_i$, all projective, then dim $S = \max_{i \in \mathbb{N}} \dim S_i$.

We replace (SP) by the weaker property:

(WP) **Weak properness**.

Given irreducible $S \subseteq_{cl} U \subseteq_{op} M^n$ and $F \subseteq_{cl} V \subseteq_{op} M^{n+k}$ with the projection pr $: M^{n+k} \to M^n$ such that $\mathrm{pr}\, F \subseteq S$, $\dim \mathrm{pr}\, F = \dim S$, there exists $D \subseteq_{op} S$ such that $D \subseteq \mathrm{pr}\, F$.

Obviously, Noetherian Zariski structures satisfy (WP).

**Exercise 6.1.1.** *Show that (CU) in the presence of (DCC) implies both (DU) and (EU) of Section 3.1.2.*

We further postulate (AF) and (FC). The following helps to understand the dimension of projective sets.

**Lemma 6.1.2.** *Let* $P = \mathrm{pr}\, S \subseteq M^n$, *for S irreducible constructible, and* $U \subseteq_{op} M^n$ *with* $P \cap U \neq \emptyset$. *Then*

$$\dim P \cap U = \dim P.$$

*Proof.* We can write $P \cap U = \mathrm{pr}\, S' = P'$, where $S' = S \cap \mathrm{pr}^{-1} U$ is constructible irreducible, dim $S' = \dim S$ by (SI). By (FC), there is $V \subseteq_{op} M^n$ such that for all $c \in V \cap P$,

$$\dim \mathrm{pr}^{-1}(c) \cap S = \min_{a \in P} \dim \mathrm{pr}^{-1}(a) \cap S = \dim S - \dim P.$$

Note that $\mathrm{pr}^{-1} U \cap \mathrm{pr}^{-1} V \cap S \neq \emptyset$, because $S$ is irreducible. By taking $s \in \mathrm{pr}^{-1} U \cap \mathrm{pr}^{-1} V \cap S$ and $c = \mathrm{pr}\, s$ and using (FC) for $S'$, we get

$$\dim \mathrm{pr}^{-1}(c) \cap S = \min_{a \in P'} \dim \mathrm{pr}^{-1}(a) \cap S = \dim S - \dim P'.$$

So, dim $P' = \dim P$.                                                            $\square$

Another useful general fact is easy to prove using (AF).

**Exercise 6.1.3.** *Given an irreducible $F \subseteq_{cl} U \subseteq_{op} M^k$, $\dim F > 0$, there is $i \leq k$ such that for $\mathrm{pr}_i : (x_1, \ldots, x_k) \mapsto x_i$,*

$$\dim \mathrm{pr}_i F > 0.$$

### 6.1.2 Analytic subsets

**Definition 6.1.4.** A subset $S$, $S \subseteq_{cl} U \subseteq_{op} M^n$, is called **analytic** in $U$ if for every $a \in S$ there is an open $V_a \subseteq_{op} U$ such that $S \cap V_a$ is the union of finitely many relatively closed irreducible subsets.

We postulate the following properties.

(INT) **Intersections:** If $S_1$, $S_2 \subseteq_{an} U$ are irreducible, then $S_1 \cap S_2$ is analytic in $U$,

(CMP) **Components:** If $S \subseteq_{an} U$ and $a \in S$, then there is $S_a \subseteq_{an} U$, a finite union of irreducible analytic subsets of $U$, and some $S'_a \subseteq_{an} U$ such that $a \in S_a \setminus S'_a$ and $S = S_a \cup S'_a$,

Each of the irreducible subsets of $S_a$ is called an **irreducible component of $S$ (containing $a$).**

(CC) **Countability of the number of components:** Any $S \subseteq_{an} U$ is a union of at most countably many irreducible components.

**Remark 6.1.5.** It is immediate that an irreducible analytic subset is strongly irreducible. Also, it is easy to see that in a Noetherian Zariski structure, closed subsets of open sets are analytic, so the property (SI) postulated for Noetherian Zariski structures holds in analytic Zariski ones, although in a more careful formulation.

**Exercise 6.1.6.** *For $S$ analytic and $a \in \mathrm{pr}\, S$, the fibre $S(a, M)$ is analytic.*

**Lemma 6.1.7.** *If $S \subseteq_{an} U$ is irreducible, $V$ open, then $S \cap V$ is an irreducible analytic subset of $V$ and, if non-empty, $\dim S \cap V = \dim S$.*

*Proof.* Proof is immediate by (SI). $\square$

**Exercise 6.1.8.**

(i) $\emptyset$, *any singleton, and $U$ are analytic in $U$;*
(ii) *if $S_1$, $S_2 \subseteq_{an} U$, then $S_1 \cup S_2$ is analytic in $U$;*
(iii) *if $S_1 \subseteq_{an} U_1$ and $S_2 \subseteq_{an} U_2$, then $S_1 \times S_2$ is analytic in $U_1 \times U_2$;*
(iv) *if $S \subseteq_{an} U$ and $V \subseteq U$ are open, then $S \cap V \subseteq_{an} V$; and*
(v) *if $S_1$, $S_2 \subseteq_{an} U$, then $S_1 \cap S_2$ is analytic in $U$.*

**Definition 6.1.9.** Given a subset $S \subseteq_{cl} U \subseteq_{op} M^n$, we define the notion of the **analytic rank** of $S$ in $U$, $\text{ark}_U(S)$, which is a natural number satisfying

1. $\text{ark}_U(S) = 0$ iff $S = \emptyset$;
2. $\text{ark}_U(S) \leq k + 1$ iff there is a set $S' \subseteq_{cl} S$ such that $\text{ark}_U(S') \leq k$ and with the set $S^0 = S \setminus S'$ being analytic in $U \setminus S'$.

Obviously, any non-empty analytic subset of $U$ has analytic rank 1.

**Example 6.1.10.** In a previous work (Zilber, 1997), we have discussed the following notion of generalised analytic subsets of $[\mathbf{P}^1(\mathbb{C})]^n$ and, more generally, of $[\mathbf{P}^1(K)]^n$ for $K$ algebraically closed complete valued field.

Let $F \subseteq \mathbb{C}^2$ be a graph of an entire analytic function and $\bar{F}$ its closure in $[\mathbf{P}^1(\mathbb{C})]^2$. It follows from Picar's theorem that $\bar{F} = F \cup \{\infty\} \times \mathbf{P}^1(\mathbb{C})$ and in particular that $\bar{F}$ has analytic rank 2.

**Generalised analytic sets** are defined as the subsets of $[\mathbf{P}^1(\mathbb{C})]^n$ for all $n$, obtained from classical (algebraic) Zariski closed subsets of $[\mathbf{P}^1(\mathbb{C})]^n$ and $\bar{F}$ by applying the positive operations: Cartesian products, finite intersections, unions, and projections.

It has been proven (by a simple induction on the number of operation) that *any generalised analytic set is of finite analytic rank* (Zilber, 1997).

The next assumption guarantees that the class of analytic subsets explicitly determines the class of closed subsets in $M$:

(AS) **Analytic stratification:** For any $S \subseteq_{cl} U \subseteq_{op} M^n$, $\text{ark}_U S$ is defined and finite.

We also are going to consider the following property:

(PS) **Pre-smoothness:** If $S_1, S_2 \subseteq_{an} U \subseteq_{op} M^n$ are both $S_1, S_2$ irreducible, then for any irreducible component $S_0$ of $S_1 \cap S_2$

$$\dim S_0 \geq \dim S_1 + \dim S_2 - \dim U.$$

**Definition 6.1.11.** A topological structure M with good dimension satisfying axioms (INT)–(AS) is called an **analytic Zariski structure**. We also assume throughout that $M$ is irreducible. An analytic Zariski structure is called pre-smooth if it has the pre-smoothness property (PS).

# 6.2 Compact analytic Zariski structures

We consider in this section the case of a compact M. Our aim is to prove the following theorem, stressing the fact that the notion of analytic Zariski generalises the one considered in Chapter 3.

**Theorem 6.2.1.** *Let* $M = (M, C)$ *be a compact analytic Zariski structure and* $C^0$ *be the sub-family of* $C$ *consisting of subsets analytic in* $M^n$, *all* $n$. *Then* $(M, C^0)$ *is a Noetherian Zariski structure.*

This is an abstract analogue of Theorems 3.4.3 and 3.4.7 about complex and rigid analytic manifolds (including the Chow theorem). The proofs are from the work of Peatfield and Zilber (2004).

Proof of the theorem: to compare the definitions to prove the theorem, we need only to check the descending chain condition (DCC) for $C^0$ and the fact that $C^0$ is closed under projections (proper mapping theorem). This is proved in the following lemmata. □

**Lemma 6.2.2.** *Analytic subsets of* $M^n$ *have only finitely many irreducible components.*

*Proof.* Suppose $S \subseteq_{an} M^n$ has infinitely many components. Then by (CMP), for any $a \in S$, we have a closed subset $S'_a \subseteq S$ which does not contain $a$ and contains all but finitely many components of $S$.

Obviously, the family $\{S'_a : a \in S\}$ is filtering. Thus, by compactness, there must be a common point for all members of the family, which is a contradiction. □

**Lemma 6.2.3.** $C^0$ *satisfies (DCC).*

*Proof.* By finiteness, dimension stabilises in any descending $C^0$-chain. By Lemma 6.2.2, the chain stabilises. □

**Lemma 6.2.4.** *For any* $S \in C^0$, *we have* $\mathrm{pr}\, S \in C^0$.

*Proof.* We may assume that $S$ is irreducible. Then $\mathrm{pr}\, S$ is closed in $M^m$ by compactness and cannot be represented as a non-trivial union $R_1 \cup R_2$ of two closed subsets (consider inverse images of $R_1$ and $R_2$ in $S$). By definition, $\mathrm{pr}\, S$ is analytic in $M^m$.

This finishes the proof of the theorem. □

Now we concentrate on the proof of the proper mapping theorem.

**Lemma 6.2.5.** *Let* $V \subseteq M^n$ *be an open subset and*

$$\{T^b : b \in B\}$$

*be a definable family of analytic subsets* $T^b \subseteq_{an} V$. *Then*

$$T^* = \bigcap_{b \in B} T^b \subseteq_{an} V.$$

*Proof.* Suppose $a \in T^*$. Then, because finite intersections are analytic again, for any $b_1, \ldots, b_k \in B$ there are finitely many irreducible components of $T^{b_1} \cap \cdots \cap T^{b_k}$ which contain $a$. Let $(T^{b_1} \cap \cdots \cap T^{b_k})_a$ be the union of the components and choose $b_1, \ldots, b_k$, depending on $a$, so that the number of the components and the dimension of each of them are minimal possible. Then

$$(T^{b_1} \cap \cdots \cap T^{b_k})_a = (T^{b_1} \cap \cdots \cap T^{b_k})_a \cap T^*.$$

We can now find, by (CMP), a subset $(T^{b_1} \cap \cdots \cap T^{b_k})'_a$, closed in $V$, which does not contain $a$ and such that

$$(T^{b_1} \cap \cdots \cap T^{b_k})_a \cup (T^{b_1} \cap \cdots \cap T^{b_k})'_a = (T^{b_1} \cap \cdots \cap T^{b_k}).$$

Let

$$V_a = V \setminus (T^{b_1} \cap \cdots \cap T^{b_k})'_a.$$

Then

$$T^* \cap V_a = (T^{b_1} \cap \cdots \cap T^{b_k})_a \cap V_a;$$

that is, $T^*$ in the neighbourhood is equal to a finite union of irreducible sets.

If $a \notin T^*$, then there is $b \in B$ such that $a \notin T^b$. By putting $V_a = V \setminus T^b$, we have $a \in V_a$ and clearly $T^* \subseteq T^b$ so that $T^* \cap V_a = \emptyset$, the empty union of sets irreducible in $V_a$. $\square$

**Lemma 6.2.6.** *If $S \subseteq_{cl} W \subsetneq_{op} M^n$ and $C \subseteq W$ is such that $C$ is closed in $M^n$, then $C \cap S$ is closed in $M^n$.*

*Proof.* Say $S = S_c \cap W$ where $S_c \subseteq_{cl} M^n$. Then $C \cap S = C \cap S_c \cap W = C \cap S_c$ is closed in $M^n$. $\square$

**Lemma 6.2.7.** *Let $S \subseteq_{an} W \subseteq_{op} M^n$ and $C \subseteq S$ be such that $C \subseteq_{cl} M^n$. Then there are $S_1, \ldots, S_k$ such that each $S_i$ is closed and irreducible in $W$ and $S' \subseteq_{an} U$ such that $S = \bigcup_{i=1}^k S_i \cup S'$ and $C \cap S' = \emptyset$.*

*Proof.* First note that for any $a \in C$, we have $a \in S$, so by the analyticity of $S$ there is $S_a$, a finite union of sets irreducible in $W$, and $S'_a \subseteq_{an} W$ such that $S = S_a \cup S'_a$ and $a \notin S'_a$. Consider $\bigcap \{S'_a | a \in S\}$. For any $a \in C$, $a \notin S'_a$ and so $a \notin \bigcap \{S'_a | a \in S\}$. Thus, $C \cap \bigcap \{S'_a | a \in S\} = \emptyset$, that is, $\bigcap \{(C \cap S'_a) | a \in S\} = \emptyset$. Now, because $C \subseteq S$ and $C \subseteq_{cl} M^n$, we have that $C \cap S'_a \subseteq_{cl} M^n$, and then by compactness we have that there must be an empty finite sub-intersection. Say $a_1, \ldots, a_k \in S$ are such that $\bigcap_{i=1}^k (C \cap S'_{a_k}) = \emptyset$, so that by writing $S' = \bigcap_{i=1}^k S'_{a_k}$, we get $C \cap S' = \emptyset$. Also, by writing $S_i$ and $S'_i$ for $S_{a_i}$

and $S'_{a_i}$ respectively, we note $S \setminus S_i \subseteq S'_i$, and so

$$S = \bigcup_{i=1}^{k} S_i \cup (S \setminus \bigcup_{i=1}^{k} S_i) = \bigcup_{i=1}^{k} S_i \cup \bigcap_{i=1}^{k} (S \setminus S_i) \subseteq$$

$$\subseteq \bigcup_{i=1}^{k} S_i \cup \bigcap_{i=1}^{k} S'_i = \bigcup_{i=1}^{k} S_i \cup S' \subseteq S.$$

Thus, we get equality throughout. Because each $S_i$ is a finite union of sets irreducible in $W$, this gives the result. $\qquad \square$

Let $S \subseteq_{\mathrm{an}} W \subseteq_{\mathrm{op}} M^n$ and pr $: M^n \to M^m$ be the standard projection map, with $\mathrm{pr}(W) = U$, $\mathrm{pr}(S) \subseteq U \subseteq_{\mathrm{op}} M^m$. We say that the projection is **proper** on $S$ if for any irreducible component $S_i$ of $S$ we have that the pr $S_i$ is closed in $U$ and for any $a \in \mathrm{pr} \, S$ we have that the fibre $\mathrm{pr}^{-1}(a) \cap S$ is compact in $M^n$.

**Theorem 6.2.8 (proper mapping theorem).** *Given* $S \subseteq_{\mathrm{an}} W \subseteq_{\mathrm{op}} M^n$ *and* pr $: M^n \to M^m$, *a standard projection such that* $\mathrm{pr} \, S \subseteq U \subseteq_{\mathrm{op}} M^m$, *suppose* pr *is proper on* $S$. *Then* pr $S$ *is analytic in* $U$.

*Proof.* Say $a \in \mathrm{pr}(S)$ and note that by properness $S_a = \mathrm{pr}^{-1}(a) \cap S \subseteq_{\mathrm{cl}} M^n$. Also $S_a \subseteq S$, and so by the lemma there are sets, $S_1, \ldots, S_k$, irreducible in $W$ and $S' \subseteq_{\mathrm{an}} W$ such that $S_a \cap S' = \emptyset$ and $S = \bigcup_{i=1}^{k} S_i \cup S'$. Now, $S_a \cap S' = \mathrm{pr}^{-1}(a) \cap S' = \emptyset$ and so $a \notin \mathrm{pr}(S')$. Thus, by putting $U_a = U \setminus \mathrm{pr}(S') \subseteq_{\mathrm{op}} M^m$, we get $a \in U_a$. By properness, each $\mathrm{pr}(S_i)$ is closed in $U$, and because each $S_i$ is irreducible, each $\mathrm{pr}(S_i)$ is also irreducible. If it were not, then there would be some $C \subsetneq \mathrm{pr}(S_i)$ with $\dim(C) = \dim(\mathrm{pr}(S_i))$. Then by (AF) we would have

$$\dim S_i = \dim \mathrm{pr} \, S_i + \min_{a \in \mathrm{pr} \, S_i} (\dim(\mathrm{pr}^{-1}(a) \cap S_i))$$

$$\leq \dim C + \min_{a \in C}(\dim(\mathrm{pr}^{-1}(a) \cap S_i))$$

$$\leq \dim((C \times M^{n-m}) \cap S_i),$$

and because $(C \times M^{n-m}) \cap S_i \subsetneq S_i$, this contradicts the irreducibility of $S_i$. Thus

$$\mathrm{pr}(S) \cap U_a = \mathrm{pr} \left( \bigcup_{i=1}^{k} S_i \cup S' \right) \cap (U \setminus \mathrm{pr}(S'))$$

$$= \left( \bigcup_{i=1}^{k} \mathrm{pr}(S_i) \cup \mathrm{pr}(S') \right) \cap (U \setminus \mathrm{pr}(S'))$$

$$= \bigcup_{i=1}^{k} (\mathrm{pr}(S_i) \cap U_a),$$

which is a finite set of irreducibles, because closed projection of a strongly irreducible set is strongly irreducible. Thus, pr $S$ is analytic at $a$.                    □

## 6.3 Model theory of analytic Zariski structures

In contrast with the theory of Noetherian Zariski structures, the model theory of analytic Zariski structures is essentially non-elementary (non-first-order). This manifests itself, first of all, in the fact that we have to treat arbitrary infinite intersections of closed sets which presumes at least some use of $L_{\infty,\omega}$-language rather than the first-order one. Some of the properties, such as (CC), need even more powerful language, with the quantifier 'there is uncountably many $v$ such that ...'

We are going to prove here a theorem which in effect states a non-elementary quantifier elimination to the level of existential formulas, assuming that our analytic Zariski structure M is of dimension 1. Pre-smoothness is not needed. This can be seen as an analogue of Theorem 3.2.1. Another related result states that M is $\omega$-stable in the sense of *abstract elementary classes*; this is an analogue of Theorem 3.2.8. The reader can also see the relevance of Hrushovski's pre-dimension arguments in this context (see Section B.2.2).

**Definition 6.3.1.** Let $M_0$ be a non-empty subset of $M$ and $C_0$, a sub-family of $C$. We say that $(M_0, C_0)$ is a **core sub-structure** if

1. for each $\{\langle x_1, \ldots, x_n \rangle\} \in C_0$ (a singleton), $x_1, \ldots, x_n \in M_0$;
2. $C_0$ satisfies (L1)–(L6) (Section 2.1) and (L7) with $a \in M_0^k$;
3. $C_0$ satisfies (WP), (AF), (FC), and (AS);
4. for any $C_0$-constructible $S \subseteq_{an} U \subseteq_{op} M^n$, every irreducible component $S_i$ of $S$ is $C_0$-constructible; and
5. for any non-empty $C_0$-constructible $U \subseteq M$, $U \cap M_0 \neq \emptyset$.

**Exercise 6.3.2.** *Given any countable $N \subseteq M$ and $C \subseteq C$, there exist countable $M_0 \supseteq N$ and $C_0 \supseteq C$ such that $(M_0, C_0)$ is a sub-model.*

We fix a core sub-structure $(M_0, C_0)$ with $M_0$ and $C_0$ countable.

**Definition 6.3.3.** For finite $X \subseteq M$, we define the $C_0$-**pre-dimension**

$$\delta(X) = \min\{\dim S : \vec{X} \in S, \ S \subseteq_{an} U \subseteq_{op} M^n, \ S \text{ is } C_0\text{-constructible}\}$$

and **dimension**

$$\partial(X) = \min\{\delta(XY) : Y \subseteq M\}.$$

For $X \subseteq M$ finite, we say that $X$ is **self-sufficient** and write $X \leq M$ if $\partial(X) = \delta(X)$.

For infinite $A \subseteq M$, we say $A \leq M$ if for any finite $X \subseteq A$ there is a finite $X \subseteq X' \subseteq A$ such that $X' \leq M$.

We work now under the assumption that dim $M = 1$. Note that we then have

$$0 \leq \delta(Xy) \leq \delta(X) + 1, \text{ for any } y \in M,$$

because $\vec{X}y \in S \times M$.

**Proposition 6.3.4.** *Let* $P = \text{pr } S$, *for some* $C_0$-*constructible* $S \subseteq_{an} U \subseteq_{op} M^{n+k}$, pr : $M^{n+k} \to M^n$. *Then*

$$\dim P = \max\{\partial(x) : x \in P(M)\}. \tag{6.1}$$

*Moreover, this formula is true when* $S \subseteq_{cl} U \subseteq_{op} M^{n+k}$.

*Proof.* We use induction on dim $S$.

We first note that by induction on $\text{ark}_U S$, if Eq. (6.1) holds for all analytic $S$ of dimension less or equal to $k$, then it holds for all closed $S$ of dimension less or equal to $k$.

The statement is obvious for dim $S = 0$, and so we assume that dim $S > 0$ and that for all analytic $S'$ of lower dimension the statement is true.

By (CU) and (CMP), we may assume that $S$ is irreducible. Then by (AF)

$$\dim P = \dim S - \dim S(c, M) \tag{6.2}$$

for any $c \in P(M) \cap V(M)$ [such that $S(c, M)$ is of minimal dimension] for some open $C_0$-constructible $V$.

*Claim 1.* It is enough to prove the statement of the proposition for the projective set $P \cap V'$, for some $C_0$-open $V' \subseteq_{op} M^n$.

Indeed,

$$P \cap V' = \text{pr}(S \cap \text{pr}^{-1}V'), \qquad S \cap \text{pr}^{-1}V' \subseteq_{cl} \text{pr}^{-1}V' \cap U \subseteq_{op} M^{n+k}$$

and $P \setminus V' = \text{pr}(S \cap T)$, $T = \text{pr}^{-1}(M^n \setminus V') \in C_0$. So, $P \setminus V'$ is the projection of a proper analytic subset of lower dimension. By induction, for $x \in P \setminus V'$, $\partial(x) \leq \dim P \setminus V' \leq \dim P$, and hence, using Lemma 6.1.2,

$$\dim P \cap V' = \max\{\partial(x) : x \in P \cap V'\} \Rightarrow \dim P = \max\{\partial(x) : x \in P\}.$$

*Claim 2.* The statement of the proposition holds if dim $S(c, M) = 0$ in Equation (6.2).

*Proof.* Given $x \in P$, choose a tuple $y \in M^k$ such that $S(x^\frown y)$ holds. Then $\delta(x^\frown y) \leq \dim S$, so we have $\partial(x) \leq \delta(x^\frown y) \leq \dim S = \dim P$.

It remains to notice that there exists $x \in P$ such that $\partial(x) \geq \dim P$. Consider the $C_0$-type

$$x \in P \ \& \ \{x \notin R : \dim R \cap P < \dim P \text{ and } R \text{ is projective}\}.$$

This is realised in $M$, because otherwise $P = \bigcup_R (P \cap R)$ which would contradict (CU) because $(M_0, C_0)$ is countable.

For such an $x$, let $y$ be a tuple in $M$ such that $\delta(x^\frown y) = \partial(x)$. By definition, there exist $S' \subseteq_{an} U' \subseteq_{op} M^m$ such that $\dim S' = \delta(x^\frown y)$. Let $P' = \mathrm{pr}\, S'$, the projection into $M^n$. By our choice of $x$, $\dim P' \geq \dim P$, but $\dim S' \geq \dim P'$. Hence, $\partial(x) \geq \dim P$. The claim is proved.

*Claim 3.* There is a $C_0$-constructible $R \subseteq_{an} S$ such that all the fibres $R(c, M)$ of the projection map $R \to \mathrm{pr}\, R$ are 0-dimensional and $\dim \mathrm{pr}\, R = \dim P$.

*Proof.* We have by construction $S(c, M) \subseteq M^k$. Assuming $\dim S(c, M) > 0$ on every open subset, we show that there is a $b \in M_0$ such that (up to the order of coordinates) $\dim S(c, M) \cap \{b\} \times M^{k-1} < \dim S(c, M)$, for all $c \in P \cap V' \neq \emptyset$, for some open $V' \subseteq V$, and $\dim \mathrm{pr}\, S(c, M) \cap \{b\} \times M^{k-1} = \dim P$. By induction on $\dim S$, this proves the claim.

To find such a $b$, choose $a \in P \cap V$ and note that by Exercise 6.1.3, up to the order of coordinates, $\dim \mathrm{pr}_1 S(a, M) > 0$, where $\mathrm{pr}_1 : M^k \to M$ is the projection on the first co-ordinate.

Consider the projection $\mathrm{pr}_{M^n, 1} : M^{n+k} \to M^{n+1}$ and the set $\mathrm{pr}_{M^n, 1} S$. By (AF), we have

$$\dim \mathrm{pr}_{M^n, 1} S = \dim P + \dim \mathrm{pr}_1 S(a, M) = \dim P + 1.$$

By using (AF) again for the projection $\mathrm{pr}^1 : M^{n+1} \to M$ with the fibres $M^n \times \{b\}$, we get, for all $b$ in some open subset of $M$,

$$1 \geq \dim \mathrm{pr}^1 \mathrm{pr}_{M^n, 1} S = \dim \mathrm{pr}_{M^n, 1} S - \dim [\mathrm{pr}_{M^n, 1} S] \cap [M^n \times \{b\}]$$
$$= \dim P + 1 - \dim [\mathrm{pr}_{M^n, 1} S] \cap [M^n \times \{b\}].$$

Hence, $\dim [\mathrm{pr}_{M^n, 1} S] \cap [M^n \times \{b\}] \geq \dim P$, for all such $b$, which means that the projection of the set $S_b = S \cap (M^n \times \{b\} \times M^{k-1})$ on $M^n$ is of dimension $\dim P$, which finishes the proof if $b \in M_0$, but $\dim S_b = \dim S - 1$ for all $b \in M \cap V'$, some $C_0$-open $V'$, so for any $b \in M_0 \cap V'$. The latter is not empty because $(M_0, C_0)$ is a sub-model. This proves the claim.

*Claim 4.* Given $R$ satisfying claim 3,

$$P \setminus \mathrm{pr}\, R \subseteq \mathrm{pr}\, S', \qquad \text{for some } S' \subseteq_{cl} S, \ \dim S' < \dim S.$$

*Proof.* Consider the Cartesian power

$$M^{n+2k} = \{x^\frown y^\frown z : x \in M^n, \ y \in M^k, \ z \in M^k\}$$

and its $C_0$-constructible subset

$$R\&S := \{x^\frown y^\frown z : \ x^\frown z \in R \ \& \ x^\frown y \in S\}.$$

Clearly $R\&S \subseteq_{an} W \subseteq_{op} M^{n+2k}$, for an appropriate $C_0$-constructible $W$.

Now notice that the fibres of the projection $\mathrm{pr}_{xy} : x^\frown y^\frown z \mapsto x^\frown y$ over $\mathrm{pr}_{xy} R\&S$ are 0-dimensional and so, for some irreducible component $(R\&S)^0$ of the analytic set $R\&S$, $\dim \mathrm{pr}_{xy}(R\&S)^0 = \dim S$. Because $\mathrm{pr}_{xy} R\&S \subseteq S$ and $S$ irreducible, we get by (WP) $D \subseteq \mathrm{pr}_{xy} R\&S$ for some $D \subseteq_{op} S$. Clearly

$$\mathrm{pr}\, R = \mathrm{pr}\, \mathrm{pr}_{xy} R\&S \supseteq \mathrm{pr}\, D,$$

and $S' = S \setminus D$ satisfies the requirement of the claim.

Now we complete the proof of the proposition: by claims 2 and 3

$$\dim P = \max_{x \in \mathrm{pr}\, R} \partial(x).$$

By induction on $\dim S$, using claim 4, for all $x \in P \setminus \mathrm{pr}\, R$,

$$\partial(x) \leq \dim \mathrm{pr}\, S' \leq \dim P.$$

The statement of the proposition follows. $\qquad\qquad\qquad\qquad\qquad\Box$

Recall the standard model-theoretic definition.

**Definition 6.3.5.** A $L_{\infty,\omega}(C_0)$-formula is constructed from the basic relations and constants corresponding to sets and singletons of $C_0$ using the following rules:

(i) for any collection of $L_{\infty,\omega}(C_0)$-formulas $\psi_\alpha(x_1, \ldots, x_n)$ (the only free variables), $\alpha \in I$, the formulas

$$\bigwedge_\alpha \psi_\alpha(x_1, \ldots, x_n) \qquad \text{and} \qquad \bigvee_\alpha \psi_\alpha(x_1, \ldots, x_n)$$

are $L_{\infty,\omega}(C_0)$-formulas; and

(ii) for any $L_{\infty,\omega}(C_0)$-formula $\psi_\alpha(x_1, \ldots, x_n)$,

$$\neg\psi_\alpha(x_1, \ldots, x_n), \quad \exists x_n\, \psi_\alpha(x_1, \ldots, x_n), \quad \text{and} \quad \forall x_n\, \psi_\alpha(x_1, \ldots, x_n)$$

are $L_{\infty,\omega}(C_0)$-formulas.

Given $\langle a_1, \ldots, a_n \rangle \in M^n$, an $L_{\infty,\omega}(C_0)$-type is the set of all $L_{\infty,\omega}(C_0)$-formulas in variables $x_1, \ldots, x_n$ which hold in M for $x_i = a_i$.

**Definition 6.3.6.** For $a \in M^n$, the **projective type of** $a$ **over** $M$ is

$$\{P(x) : a \in P, \; P \text{ is a projective set over } \mathcal{C}_0\}$$

$$\cup \{\neg P(x) : a \notin P, \; P \text{ is a projective set over } \mathcal{C}_0\}.$$

**Lemma 6.3.7.** *Suppose* $X \leq M$, $X' \leq M$, *and the (first-order) quantifier-free* $\mathcal{C}_0$*-type of* $X$ *is equal to that of* $X'$. *Then the* $L_{\infty,\omega}(\mathcal{C}_0)$*-types of* $X$ *and* $X'$ *are equal.*

*Proof.* We are going to construct a back-and-forth system for $X$ and $X'$ (see Remark A.4.23).

Let $S_X \subseteq_{\mathrm{an}} V \subseteq_{\mathrm{op}} M^n$, $S_X$ irreducible, all $\mathcal{C}_0$-constructible, and such that $X \in S_X(M)$ and $\dim S_X = \delta(X)$.

*Claim 1.* The quantifier-free $\mathcal{C}_0$-type of $X$ (and $X'$) is determined by formulas equivalent to $S_X \cap V'$, for $V'$ open such that $X \in V'(M)$.

*Proof.* Use the stratification of closed sets (AS) to choose $\mathcal{C}_0$-constructible $S \subseteq_{\mathrm{cl}} U \subseteq_{\mathrm{op}} M^n$ such that $X \in S$ and $\mathrm{ark}_U S$ is minimal. Obviously then $\mathrm{ark}_U S = 0$; that is, $S \subseteq_{\mathrm{an}} U \subseteq_{\mathrm{op}} M^n$. Now $S$ can be decomposed into irreducible components, so we may choose $S$ to be irreducible. Among all such $S$, choose one which is of minimal possible dimension. Obviously $\dim S = \dim S_X$; that is, we may assume that $S = S_X$. Now clearly any constructible set $S' \subseteq_{\mathrm{cl}} U' \subseteq_{\mathrm{op}} M^n$ containing $X$ must satisfy $\dim S' \cap S_X \geq \dim S_X$, and this condition is also sufficient for $X \in S'$.

Let $y$ be an element of $M$. We want to find a finite $Y$ containing $y$ and $Y'$ such that the quantifier-free type of $XY$ is equal to that of $X'Y'$ and both are self-sufficient in $M$. This, of course, extends the partial isomorphism $X \to X'$ to $XY \to X'Y'$ and proves the lemma.

We choose $Y$ to be a minimal set containing $y$ and such that $\delta(XY)$ is also minimal; that is,

$$1 + \delta(X) \geq \delta(Xy) \geq \delta(XY) = \partial(XY)$$

and $XY \leq M$.

We have two cases: $\delta(XY) = \partial(X) + 1$ and $\delta(XY) = \partial(X)$. In the first case, $Y = \{y\}$. By claim 1, the quantifier-free $\mathcal{C}_0$-type $r_{Xy}$ of $Xy$ is determined by the formulas of the form $(S_X \times M) \setminus T$, $T \subseteq_{\mathrm{cl}} M^{n+1}$, $T \in \mathcal{C}_0$, $\dim T < \dim(S_X \times M)$.

Consider

$$r_{Xy}(X', M) = \{z \in M : X'z \in (S_X \times M) \setminus T, \; \dim T < \dim S_X, \text{ all } T\}.$$

We claim that $r_{Xy}(X', M) \neq \emptyset$. Indeed, otherwise $M$ is the union of countably many sets of the form $T(X', M)$, but the fibres $T(X', M)$ of $T$ are of dimension 0 (because otherwise dim $T = \dim S_X + 1$, contradicting the definition of the $T$). This is impossible, by (CU).

Now we choose $y' \in r_{Xy}(X', M)$, and this is as required.

In the second case, by definition, there is an irreducible $R \subseteq_{an} U \subseteq_{op} M^{n+k}$, $n = |X|, k = |Y|$, such that $XY \in R(M)$ and dim $R = \delta(XY) = \partial(X)$. We may assume $U \subseteq V \times M^k$.

Let $P = \text{pr } R$, the projection into $M^n$. Then dim $P \le \dim R$, but also dim $P \ge \partial(X)$, by Proposition 6.3.4. Hence, dim $R = \dim P$. On the other hand, $P \subseteq S_X$ and dim $S_X = \delta(X) = \dim P$. By axiom (WP), we have $S_X \cap V' \subseteq P$ for some $C_0$-constructible open $V'$.

Hence, $X' \in S_X \cap V' \subseteq P(M)$, for $P$ the projection of an irreducible analytic set $R$ in the $C_0$-type of $XY$. By claim 1, the quantifier-free $C_0$-type of $XY$ is of the form

$$r_{XY} = \{R \setminus T : T \subseteq_{cl} R, \dim T < \dim R\}.$$

Consider

$$r_{XY}(X', M) = \{Z \in M^k : X'Z \in R \setminus T, T \subseteq_{cl} R, \dim T < \dim R\}.$$

We claim again that $r_{XY}(X', M) \ne \emptyset$.

Otherwise, the set $R(X', M) = \{X'Z : R(X'Z)\}$ is the union of countably many subsets of the form $T(X', M)$, but dim $T(X', M) < \dim R(X', M)$, as before, by (AF).

Again, $Y' \in r_{XY}(X', M)$ is as required. $\qquad\qquad\square$

**Corollary 6.3.8.** *There is countably many $L_{\infty,\omega}(C_0)$-types of tuples $X \le M$.*

Indeed, any such type is determined uniquely by the choice of a $C_0$-constructible $S_X \subseteq_{an} U \subseteq_{op} M^n$ such that dim $S_X = \partial(X)$.

**Lemma 6.3.9.** *Suppose, for finite $X, X' \subseteq M$, the projective $C_0$-types of $X$ and $X'$ coincide. Then the $L_{\infty,\omega}(C_0)$-types of the tuples are equal.*

*Proof.* Choose finite $Y$ such that $\partial(X) = \delta(XY)$. Then $XY \le M$. Let $XY \in S \subseteq_{an} U \subseteq_{op} M^n$ be $C_0$-constructible and such that dim $S$ is minimal possible; that is, dim $S = \delta(XY)$. We may assume that $S$ is irreducible. Notice that for every proper closed $C_0$-constructible $T \subseteq_{cl} U$, $XY \notin T$ by dimension considerations.

By assumptions of the lemma $X'Y' \in S$, for some $Y'$ in $M$, we also have $X'Y' \notin T$, for any $T$ as before, because otherwise a projective formula would imply that $XY'' \in T$ for some $Y''$, contradicting that $\partial(X) > \dim T$.

We also have $\delta(X'Y') = \dim S$, but for no finite $Z'$ is it possible that $\delta(X'Z') < \dim S$, for then again a projective formula would imply that $\delta(XZ) < \dim S$, for some $Z$.

It follows that $X'Y' \leq M$ and the quantifier-free types of $XY$ and $X'Y'$ coincide; hence the $L_{\infty,\omega}(C_0)$-types are equal, by Lemma 6.3.7.  $\square$

**Definition 6.3.10.** Set, for finite $X \subseteq M$,

$$\mathrm{cl}_{C_0}(X) = \{y \in M : \partial(Xy) = \partial(X)\}.$$

We fix $C_0$ and omit the subscript.

**Lemma 6.3.11.** *Then* $b \in \mathrm{cl}(A)$, *for* $\vec{A} \in M^n$, *iff* $b \in P(\vec{A}, M)$ *for some projective* $P \subseteq M^{n+1}$ *such that* $P(\vec{A}, M)$ *is at most countable. In particular,* $\mathrm{cl}(A)$ *is countable for any finite* $A$.

*Proof.* Let $d = \partial(A) = \delta(AV)$, and $\delta(AV)$ is minimal for all possible finite $V \subseteq M$. By definition $d = \dim S_0$, some analytic irreducible $S_0$ such that $\vec{A}V \in S_0$ and $S_0$ of minimal dimension. This corresponds to a $C_0$-definable relation $S_0(x, v)$, where $x, v$ strings of variables are of length $n, m$.

First assume that $b$ belongs to a countable $P(\vec{A}, M)$. By definition

$$P(x, y) \equiv \exists w\, S(x, y, w),$$

for some analytic $S \subseteq M^{n+1+k}$, $x, y, w$ strings of variables of length $n, 1$, and $k$ and the fibre $S(\vec{A}, b, M^k)$ is non-empty. We also assume that $P$ and $S$ are of minimal dimension, answering this description. By (FC), (AS), and minimality, we may choose $S$ so that $\dim S(\vec{A}, b, M^k)$ is minimal among all the fibres $S(\vec{A}', b', M^k)$.

Consider the analytic set $S^\sharp \subseteq_{\mathrm{an}} U \subseteq_{\mathrm{op}} M^{n+m+1+k}$ given by $S_0(x, v)\,\&\, S(x, y, w)$. By (AF), considering the projection of the set on $(x, v)$-co-ordinates,

$$\dim S^\sharp \leq \dim S_0 + \dim S(\vec{A}, M, M^k),$$

because $S(\vec{A}, M, M^k)$ is a fibre of the projection. Now we note that by countability $\dim S(\vec{A}, M, M^k) = \dim S(\vec{A}, b, M^k)$, so

$$\dim S^\sharp \leq \dim S_0 + \dim S(\vec{A}, b, M^k).$$

Now the projection $\mathrm{pr}_w S^\sharp$ along $w$ (corresponding to $\exists w\, S^\sharp$) has fibres of the form $S(\vec{X}, y, M^k)$, so by (AF)

$$\dim \mathrm{pr}_w S^\sharp \leq \dim S_0 = d.$$

Projecting further along $v$, we get $\dim \mathrm{pr}_v \mathrm{pr}_w S^\sharp \leq d$, but $\vec{A}b \in \mathrm{pr}_v \mathrm{pr}_w S^\sharp$, so by Proposition 6.3.4 $\partial(\vec{A}b) \leq d$. The inverse inequality holds by definition, so the equality holds. This proves that $b \in \mathrm{cl}(A)$.

Now, for the converse, we assume that $b \in \mathrm{cl}(A)$. So, $\partial(\vec{A}b) = \partial(\vec{A}) = d$. By definition, there is a projective set $P$ containing $\vec{A}b$, defined by the formula $\exists w\, S(x, y, w)$ for some analytic $S$, $\dim S = d$. Now $\vec{A}$ belongs to the projective set $\mathrm{pr}_y P$ (defined by the formula $\exists y \exists w\, S(x, y, w)$) so by Proposition 6.3.4 $d \leq \dim \mathrm{pr}_y P$, but $\dim \mathrm{pr}_y P \leq \dim P \leq \dim S = d$. Hence, all the dimensions are equal and so the dimension of the generic fibre is 0. As before, we may assume without loss of generality that all fibres are of minimal dimension, so

$$\dim S(\vec{A}, M, M^k) = 0.$$

Hence, $b$ belongs to a 0-dimensional set $\exists w\, S(\vec{A}, y, w)$, which is projective and countable. $\qquad \Box$

**Lemma 6.3.12.**

*(i)*

$$\mathrm{cl}(\emptyset) = \mathrm{cl}(M_0) = M_0.$$

*(ii) Given finite $X \subseteq M$, $y, z \in M$,*

$$z \in \mathrm{cl}(X, y) \setminus \mathrm{cl}(X) \Rightarrow y \in \mathrm{cl}(X, z).$$

*(iii)*

$$\mathrm{cl}(\mathrm{cl}(X)) = \mathrm{cl}(X).$$

*Proof.*

(i) Clearly $M_0 \subseteq \mathrm{cl}(\emptyset)$, by definition.
   We need to show the converse; that is, if $\partial(y) = 0$, for $y \in M$, then $y \in M_0$. By definition, $\partial(y) = \partial(\emptyset) = \min\{\delta(Y) : y \in Y \subset M\} = 0$. So, $y \in Y$, $\vec{Y} \in S \subseteq_{\mathrm{an}} U \subseteq_{\mathrm{op}} M^n$, $\dim S = 0$. The irreducible components of $S$ are points (singletons), so $\{\vec{Y}\}$ is one and must be in $\mathcal{C}_0$, because $(M_0, \mathcal{C}_0)$ is a core sub-structure. By Definition 6.3.1.1, $y \in M_0$.

(ii) Assuming the left-hand side $\partial(Xyz) = \partial(Xy) > \partial(X)$, $\partial(Xz) > \partial(X)$, then, by the definition of $\partial$,

$$\partial(Xy) = \partial(X) + 1 = \partial(Xz),$$

   so $\partial(Xzy) = \partial(Xz)$, $y \in \mathrm{cl}(Xz)$.

(iii) This is immediate by Lemma 6.3.11. $\qquad \Box$

By summarising this, we get the following.

**Theorem 6.3.13.**

(i) *Every $L_{\infty,\omega}(C_0)$-type realised in* M *is equivalent to a projective type, that is, a type consisting of existential (first-order) formulas and the negations of existential formulas.*

(ii) *There are only countably many $L_{\infty,\omega}(C_0)$-types realised in* M.

(iii) $(M, \mathrm{cl})$ *is a pre-geometry satisfying the countable closure property in the sense of Sections B.1.1 and B.2.2.*

*Proof.*

(i) This is immediate from Lemma 6.3.9.

(ii) By Corollary 6.3.8, there are only countably many types of finite tuples $Z \leq M$. Let $N \subseteq M_0$ be a countable subset of $M$ such that any finite $Z \leq M$ is $L_{\infty,\omega}(C_0)$-equivalent to some tuple in $N$. Every finite tuple $X \subset M$ can be extended to $XY \leq M$, so there is a $L_{\infty,\omega}(C_0)$-monomorphism $XY \to N$. This monomorphism identifies the $L_{\infty,\omega}(C_0)$-type of $X$ with one of a tuple in $N$; hence, there are no more than countably many such types.

(iii) This is proved by Lemma 6.3.12.                                    □

Note the connection of (i) to the property (EC) of Hrushovski's construction discussed in Section B.2.2. Indeed, if M were saturated, (i) would imply that every formula is equivalent to a Boolean combination of existential formulas. This is weaker than the quantifier elimination statement proved in Theorem 3.2.1 for Noetherian Zariski structures but reflects adequately the more complex nature of the notion of analytic Zariski.

Claim (ii) effectively states the non-elementary $\omega$-stability of the class $\mathcal{M}$ of sub-models of M treated as an *abstract elementary class* [see works by Grossberg (2002) and Baldwin (forthcoming) for the general theory]. The analogous statement for Noetherian Zariski structures is Theorem 3.2.8. In addition to this, we showed that the class $\mathcal{M}$ is *quasi-minimal $\omega$-homogeneous* and, provided the class also satisfies the assumption of *excellence*, $\mathcal{M}$ can be canonically extended so that in every uncountable cardinality $\kappa$ there is a unique, up-to-isomorphism, analytic Zariski $C_0$-structure M′ extending M (Zilber, 2008a). This would be an analogue of Theorem 3.5.25 for Noetherian Zariski structures, if excellence of $\mathcal{M}$ were known. It is tempting to conjecture that it holds for any analytic Zariski M, for large enough $C_0$. This is true for all known examples; see some of these in Section 6.5. A proper discussion of the assumption of excellence requires more model-theoretic work, and we skip it here. A previous paper (Zilber, 2002b) contains an analysis of this condition for a particular class of analytic Zariski structures – universal covers of semi-Abelian varieties. We have shown that in this class the condition is equivalent to certain arithmetic conjectures in the theory of semi-Abelian varieties. See more discussion of this in Section 6.5.

Claim (iii) is an important model-theoretic property. The closure operator cl is an analogue of *algebraic closure* (algebraic dependence) in algebraic geometry. The notion of the pre-dimension $\delta$ and the related (combinatorial) dimension $\partial$ is a prominent ingredient in Hrushovski's construction (see Sections B.2.2 and also B.1.1) and can be seen as relating the more complex theory of analytic dimension to the theory of dimension in algebraic geometry.

## 6.4 Specialisations in analytic Zariski structures

We work here in the natural language of topological structures given by (L), and $^*M \succ M$ is an elementary extension in this language. We study universal specialisations (Definition 2.2.12) $\pi : {}^*M \to M$ over an analytic Zariski structure M. Recall that for $a \in M^n$, the notation $\mathcal{V}_a$ stands for an infinitesimal neighbourhood of $a$, $a \in \mathcal{V}_a \subset {}^*M^n$.

**Exercise 6.4.1.** *Lemma 3.5.13 and Corollary 3.5.14 are valid in the present context.*

**Lemma 6.4.2.** *Let $P \subseteq U \subseteq_{op} M^n$ be a projective subset with* $\dim P < \dim U$. *Then, for every $a \in P$, there is an $\alpha \in \mathcal{V}_a \cap U(^*M) \setminus S(^*M)$.*

*Proof.* By Lemma 2.2.21, we just need to show that $\neg P(y) \,\&\, U(y)$ is consistent with $\mathrm{Nbd}_a(y)$. Suppose, towards a contradiction, it is not. Then

$$^*M \vDash \forall y \left( \neg P(y) \,\&\, U(y) \to Q(y, c') \right),$$

for some closed relation $Q$ and $c'$ in $^*M$ such that $a \notin Q(M, c)$, for $c = \pi(c')$. Then, for every $b \in U(M) \setminus P(M)$, $Q(b, c')$ holds. By applying $\pi$, we get $M \vDash Q(b, c)$. This proves that $U \setminus P \subseteq Q(M, c) \cap U$. By calculating dimensions, we conclude that $\dim Q(M, c) \cap U = \dim U$, but $U$ is irreducible (because $M^n$ is), so $U \subseteq Q(M, c)$. This contradicts the assumption that $a \notin Q(M, c)$. $\square$

**Exercise 6.4.3.** *Let $S \subseteq_{an} U \subseteq_{op} M^n$ be an analytic subset. Show that for any $a \in S$, if $\dim S = 0$ then*

$$\mathcal{V}_a \cap S(^*M) = \{a\}.$$

*In the general case, $V_a$ intersects only finitely many irreducible components of S.*

**Definition 6.4.4.** Given $a' \in V_a$, we define an **analytic locus of** $a'$ to be $S$, an analytic subset $S \subseteq_{an} U \subseteq_{op} M^n$ such that $a \in S$, $a' \in S(*M)$, and $S$ is of minimal possible dimension.

**Lemma 6.4.5.** *We can choose an analytic locus to be irreducible.*

*Proof.* Without loss of generality, $S$ has infinite number of irreducible components. We have

$$S(M) = \cup_{i \in \mathbb{N}} S_i(M).$$

More generally consider

$$S^k(M) = \bigcup_{i \geq k} S_i(M),$$

which are analytic subsets of $U$ by (CMP). We want to show that $a' \in S_i(*M)$ for some $i$. Otherwise, $a' \in S^k(*M)$ for all $k \in \mathbb{N}$. By definitions,

$$a \in \bigcap_{k \in \mathbb{N}} S^k(M) = \emptyset.$$

The contradiction proves the lemma.                                        □

We prove a version of Proposition 3.6.2.

**Proposition 6.4.6.** *Let $D \subseteq_{an} U \subseteq_{op} M^m$ be an irreducible set in a strongly pre-smooth M and $F \subseteq_{an} D \times V$ be an irreducible covering of $D$ discrete at $a \in D$, $V \subseteq_{op} M^k$. Then, for every $a' \in V_a \cap D(*M)$, there exists $b' \in V_b$ such that $\langle a', b' \rangle \in F$.*

*Proof.* Consider the type over $*M$,

$$p(y) = \{F(a', y)\} \cup \text{Nbd}_b(y).$$

*Claim.* $p$ is consistent.

*Proof of claim.* For closed $Q(z, y)$ and $c' \in {}^*M^n$ such that $\pi(c') = c$ and $\neg Q(c, b)$ holds, we want to find $b'$ such that $F(a', b') \& \neg Q(c', b')$. Let $L \subseteq_{an} T \subseteq_{op} D \times M^n$ be an analytic locus of $\langle a', c' \rangle$. Let

$$W = (T \times M^k) \cap \{\langle x, z, y \rangle : \neg Q(z, y)\}.$$

This is an open subset of the irreducible set $T \times M^k$; in particular,

$$\dim W = \dim T + \dim M^k = \dim D + \dim M^n + \dim M^k.$$

Now, consider an irreducible component $S$ of

$$\{\langle x, z, y \rangle \in W : F(x, y) \& L(x, z)\}$$

containing $\langle a, c, b \rangle$. By pre-smoothness,

$$\dim S \geq \dim(F \times M^n) + \dim(L \times M^k) - \dim(D \times M^{n+k}) = \dim L.$$

[Observe that $\dim F = \dim D$ by (AF).] Also, for pr : $\langle x, y, z \rangle \mapsto \langle x, z \rangle$, $\dim \text{pr} \, S = \dim S$ and so $\dim \text{pr} \, S = \dim L$. Hence by (WP), $L = \text{pr} \, S \cup L_0$ for some proper subset $L_0 \subseteq_{\text{cl}} L$, but $L$ is irreducible, so $\dim L_0 < \dim L$. We claim that $\langle a', c' \rangle \notin L_0(^*M)$. Indeed, it is immediate if $L_0$ is analytic, by the choice of $L$, but we may assume it is analytic because by axiom (AS) $L_0 = L_0^0 \cup L_0'$, for some $L' \subseteq_{\text{an}} U$ and $L_0^0 \subseteq_{\text{an}} U \setminus L_0'$. Hence $\langle a', c' \rangle \in \text{pr} \, S(^*M)$. This means that there is $b' \in {}^*M$ such that $\langle a', c' \rangle \in W(^*M)$ and $F(a', b') \& L(a', c')$ holds. This proves the claim. By Lemma 2.2.21, the proposition follows. $\qquad\qquad\qquad\qquad\qquad\qquad\qquad\qquad\qquad\qquad\qquad\qquad\square$

**Corollary 6.4.7.** *Assuming that M is strongly pre-smooth and D analytic, any function $f : D \to M$ with closed irreducible graph is strongly continuous.*

## 6.5 Examples

### 6.5.1 Covers of algebraic varieties

We consider the **universal cover of** $\mathbb{C}^\times$ as a topological structure and show that this is analytic Zariski.

This is a structure with the universe $V$ identified with the set of complex numbers $\mathbb{C}$, and we use the additive structure on it. We also consider the usual exponentiation map

$$\exp : V \to \mathbb{C}^\times$$

and want to take into our language and topology the usual Zariski topology (of algebraic varieties) on $(\mathbb{C}^\times)^n$ as well as exp as a continuous map.

A model-theoretic analysis of this structure was carried out in Zilber (2002b, 2006), Bays and Zilber (2007), and in the DPhil thesis of Lucy Smith (2008). The latter work used Zilber 2006 to provide the description of the topology $\mathcal{C}$ on $V$ and proves that $(V, \mathcal{C})$ is analytic Zariski. (It then addresses the issue of possible *compacifications* of the structure.)

Note that the whole analysis up to Corollary 6.5.9 uses only the first-order theory of the structure $(V, \mathcal{C})$, so one can wonder what changes if we replace $\mathbb{C}$ and exp with its abstract analogues. The answer to this question is known in the form of the categoricity theorem proved in our work (Zilber, 2005b, 2006) (see some corrections in Bays and Zilber, 2007): *if* ex : $U \to K^\times$ *is a group homomorphism, U is a divisible torsion-free group, K is an algebraically closed field of characteristic* 0 *and cardinality continuum, and* ker(ex) *is cyclic, then the structure is isomorphic to the structure* $(V, C)$ *on the complex numbers.* More generally, any two covers of 1-tori over algebraically closed fields of characteristic 0 of the same uncountable cardinality and with cyclic kernels are isomorphic.

We follow Smith's work (2008, pp. 17–25) with modifications and omission of some technical details.

The base of the PQF-topology (positive quantifier-free) on $V$ and its Cartesian powers $V^n$ is, in short, the family of subsets of $V^n$ defined by PQF-formulae.

**Definition 6.5.1.** A PQF-closed set is defined as a finite union of sets of the form

$$L \cap m \cdot \ln W \tag{6.3}$$

where $W \subseteq (\mathbb{C}^\times)^n$ is an algebraic sub-variety and $L$ is a $\mathbb{Q}$-**linear subspace** of $V^n$ that is defined by equations of the form $m_1 x_1 + \cdots + m_n x_n = a$, $m_i \in \mathbb{Z}$, $a \in V$.

Slightly rephrasing the quantifier elimination statement proved in our 2002 paper, Corollary 2 of Section 3, we have the following.

**Lemma 6.5.2.**

(i) *Projection of a PQF-closed set is PQF-constructible, that is, a Boolean combination of PQF-closed sets.*

(ii) *The image of a constructible set under exponentiation is a Zariski-constructible (algebraic) subset of* $(\mathbb{C}^\times)^n$. *The image of the set of the form in Equation (6.3) is Zariski closed.*

The PQF$_\omega$-topology is given by closed basic sets of the form

$$\cup_{a \in I}(S + a)$$

where $S$ is of the form of Eq. (6.3) and $I$ a subset of $(\ker \exp)^n$.

We define $C$ to be the family of all PQF$_\omega$-closed sets.

**Corollary 6.5.3.** $C$ *satisfies (L).*

We assign **dimension** to a closed set of the form of Equation (6.3)

$$\dim L \cap m \cdot \ln W := \dim \exp(L \cap m \cdot \ln W)$$

using the fact that the object on the right-hand side is an algebraic variety. We extend this to an arbitrary closed set assuming (CU); that is, that the dimension of a countable union is the maximum dimension of its members. This immediately gives (DP). By using Lemma 6.5.2 we also get (WP).

For a variety $W \subseteq (\mathbb{C}^\times)^n$, consider the system of its roots

$$W^{\frac{1}{m}} = \{\langle x_1, \ldots, x_n \rangle \in (\mathbb{C}^\times)^n : \langle x_1^m, \ldots, x_n^m \rangle \in W\}.$$

Let $d_W(m)$ be the number of irreducible components of $W^{\frac{1}{m}}$. We say that the sequence $W^{\frac{1}{m}}$, $m \in \mathbb{N}$, **stops branching** if the sequence $d_W(m)$ is eventually constant.

Obviously, in case $W$ is a singleton, $W = \{\langle w_1, \ldots, w_n \rangle\} \subseteq (\mathbb{C}^\times)^n$, and the sequence $W^{\frac{1}{m}}$ does not stop branching as $d_W(m) = m^n$. This is the simplest case when $W$ is contained in a coset of a torus, namely given by the equation $\bigwedge_i x_i = 1$. Similarly, if $W$ is contained in a coset of an irreducible torus given by $k$ independent equations of the form

$$x_1^{l_{i1}} \cdot \ldots \cdot x_n^{l_{in}} = 1,$$

then $d_W(m) = m^k$ does not stop branching.

**Fact (Zilber, 2006, Theorem 2, case $n = 1$ and its corollary).** The sequence $W^{\frac{1}{m}}$ stops branching iff $W$ is not contained in a coset of a proper sub-torus of $(\mathbb{C}^\times)^k$.

**Lemma 6.5.4.** *Any irreducible closed subset of $V^n$ is of the form of Equation (6.3), for $W$ not contained in a coset of a proper torus, $m \in \mathbb{Z}$.*

In case $W$ is contained in a coset of a proper torus $T$, note that $T = \exp L$, for some $L$ a $\mathbb{Q}$-linear subspace of $V^n$. Also, there is an obvious $\mathbb{Q}$-linear isomorphism $\sigma_L : L \to V^k$, $k = \dim L$, which induces a biregular isomorphism $\sigma_T : T \to (\mathbb{C}^\times)^k$. Now $\sigma_T(W) \subseteq (\mathbb{C}^\times)^k$ is not contained in a coset of a proper torus and so $\sigma_T(W)^{\frac{1}{m}}$ stops branching.

Note that $L$ is defined up to the shift by $a \in (\ker \exp)^n$.

**Proposition 6.5.5.** *Let $W \subseteq (\mathbb{C}^\times)^n$ be an irreducible sub-variety, $T = \exp L$ the minimal coset of a torus containing $W$ and $m$ is the level where $\sigma_T(W^{\frac{1}{m}})$ stops branching. Let $\sigma_T(W_i^{\frac{1}{m}})$ be an irreducible component of $\sigma_T(W^{\frac{1}{m}})$. Then*

$$L \cap m\sigma_T^{-1}\sigma_T\left(W_i^{\frac{1}{m}}\right) \tag{6.4}$$

*is an irreducible component of $\ln W$. Moreover, any irreducible component of $\ln W$ has this form for some choice of $L$, $\exp L = T$.*

**Remark 6.5.6.**

(i) The irreducible components of the form of Equation (6.4) for distinct choices of $L$ do not intersect.

(ii) There are finitely many irreducible components of the form of Equation (6.4) for fixed $L$ and $W$.

**Remark 6.5.7.** Proposition 6.5.5 eventually provides a description of the irreducible decomposition for any set of the form of Equation (6.3) for any closed set. Indeed, the irreducible components of the set $L \cap m \cdot \ln W$ are among irreducible components of $\ln X$, for the algebraic variety $X = \exp(L \cap m \cdot \ln W)$.

**Corollary 6.5.8.** *Any closed subset of $V^n$ is analytic in $V^n$.*

Now it is easy to check that (SI), (INT), (CMP), (CC), (AS), and (PS) are satisfied.

**Corollary 6.5.9.** *The structure $(V, \mathcal{C})$ is analytic Zariski one-dimensional and pre-smooth.*

An inquisitive reader will notice that this analysis treats only *formal* notion of analyticity on the cover $\mathbb{C}$ of $\mathbb{C}^\times$ but does not address the classical one. In particular, *is the formal analytic decomposition as described by Proposition 6.5.5 the same as the actual complex analytic one?* In a private communication, F. Campana answered this question in the positive, using a co-homological argument. M. Gavrilovich proved this and a much more general statement in his thesis (2006, III.1.2) by a similar argument.

Now we look into yet another version of a cover structure which is proven to be analytic Zariski, a **cover of the one-dimensional algebraic torus over an algebraically closed field of a positive characteristic.**

Let $(V, +)$ be a divisible torsion-free Abelian group and $K$ an algebraically closed field of a positive characteristic $p$. We assume that $V$ and $K$ are both of the same uncountable cardinality. Under these assumptions, it is easy to construct a surjective homomorphism

$$\mathrm{ex} : V \to K^\times.$$

The kernel of such a homomorphism must be a subgroup which is $p$-divisible but not $q$-divisible for each $q$ coprime with $p$. One can easily construct ex so that

$$\ker \mathrm{ex} \cong \mathbb{Z}\left[\frac{1}{p}\right],$$

the additive group (which is also a ring) of rationals of the form $m/p^n$, $m, n \in \mathbb{Z}$, $n \geq 0$. In fact, in this case it is convenient to view $V$ and $\ker \mathrm{ex}$ as $\mathbb{Z}[1/p]$-modules.

In this new situation, Lemma 6.5.2 is still true, with obvious alterations, and we can use Definition 6.5.1 to introduce a topology and the family $C$ as before. The fact right before Definition 6.5.1 for $K^\times$ is proved in Bays and Zilber (2007). Hence, the corresponding versions of Proposition 6.5.5 to Corollary 6.5.9 follow.

## 6.5.2 Hard examples

These are structures which, on the one hand, have been discovered and studied with the use of Hrushovski's model-theoretic construction (See Section B.2.2), and, on the other, conjecturally coincide with classical structures playing a central role in mathematics. Some of the Zariski topology on these structures can be guessed, but it is still not a definitive description of a natural topology.

### Pseudo-exponentiation

Refer to the field with pseudo-exponentiation $K_{ex} = (K, +, \cdot, ex)$ of Sub-section B.2.3. Theorem B.2.1 states that there is a unique, up-to-isomorphism, field with pseudo-exponentiation $K_{ex} = (K, +, \cdot, ex)$ satisfying certain axioms, of any given uncountable cardinality.

For the purposes of our discussion here, we fix one, $K_{ex}$, of cardinality continuum.

Note also that categoricity implies stability, in the same sense as in Theorem 6.3.13(ii) and Theorem B.2.2 confirm the property of analytic Zariski structures proved in Theorem 6.3.13(i). Theorem 6.3.13(iii) as well as quasi-minimality and excellence are proven properties of $K_{ex}$.

Now refer to the main conjecture discussed in Sub-section B.2.3:

$$K_{ex} \cong \mathbb{C}_{exp}.$$

In connection with this, we introduce the **natural Zariski topology on** $\mathbb{C}_{exp}$. A subset $S \subseteq \mathbb{C}^n$ will be in $C_{exp}$ if $S$ is a Boolean combination of projective subsets of $\mathbb{C}_{exp}$ and $S$ is closed in the classical metric topology on $\mathbb{C}^n$.

We hope that by assuming the main conjecture it is possible to characterise (syntactically) the subsets in $C_{exp}$, that is, define the natural Zariski topology on $K_{ex}$ without referring to the metric topology on $\mathbb{C}$.

Finally we conjecture (in conjunction with the main conjecture) that $\mathbb{C}_{exp}$ is analytic Zariski with regards to the natural Zariski topology.

The apparent candidate for the notion of analytic subsets on the abstract field $K_{ex}$ with pseudo-exponentiation is the family of sets defined by systems of polynomial-exponential equations, that is, equations of the form

$$p(x_1, \ldots, x_n, e^{x_1}, \ldots, e^{x_n}, e^{e^{x_1}}, \ldots, e^{e^{x_n}}, \ldots) = 0,$$

for $p$ a polynomial over $\mathbb{C}$. This obviously is not enough. Indeed, the function

$$g(y_1, y_2) := \begin{cases} \frac{\mathrm{ex}(y_1) - \mathrm{ex}(y_2)}{y_1 - y_2} & \text{if } y_1 \neq y_2, \\ \mathrm{ex}(y_1) & \text{otherwise} \end{cases}$$

must be Zariski continuous (cf. Definition 2.2.32), but its graph, although defined by a quantifier-free formula, is not the zero set of a system of exponential polynomials.

In accordance with Definition 2.2.32, this formula introduces the notion of the derivative of the pseudo-exponentiation ex in $K_{\mathrm{ex}}$. The abstract axiomatisation of $K_{\mathrm{ex}}$ cannot distinguish between $e^x$ and $e^{cx}$ (provided $c$ is chosen so that both satisfy Schanuel's conjecture). Interestingly, if an abstract pseudo-exponentiation ex has a derivative $\mathrm{ex}'$, then $\mathrm{ex}'(x) = c \cdot \mathrm{ex}(x)$, for some constant $c$. Indeed, the graph of the function $z = g(y_1, y_2)$ is the closure of the graph of the function $z = [\mathrm{ex}(y_1) - \mathrm{ex}(y_2)]/(y_1 - y_2)$. The latter, and hence the former, is invariant under the transformation of variables

$$y_1 \mapsto y_1 + t, \qquad y_2 \mapsto y_2 + t, \qquad z \mapsto z \cdot \mathrm{ex}(t).$$

Hence, the frontier, which is the graph of $z = \mathrm{ex}'(y_1) = \mathrm{ex}'(y_2)$, is invariant under the transformation. This means that

$$\mathrm{ex}'(y + t) = \mathrm{ex}'(y) \cdot \mathrm{ex}(t).$$

It follows that

$$\mathrm{ex}'(t) = c \cdot \mathrm{ex}(t), \qquad \text{for} \qquad c = \mathrm{ex}'(0).$$

Similarly to pseudo-exponentiation, one can consider the meromorphic Weierstrass function $\mathbb{C} \to \mathbb{C}$,

$$\wp_\Omega(z) = \sum_{\omega \in \Omega} \frac{1}{(x - \omega)^2}$$

over a given lattice $\Omega$. The model theory of a 'pseudo-Weierstrass function' is less developed (see Kirby, 2008); in particular, the analogue of Theorem B.2.1 is not known, but expected.

An interesting structure definable in $K_{\mathrm{ex}}$ (and similarly in $\mathbb{C}_{\exp}$) is the **field with raising to powers**: we denote it $K^P$. Here $P$ is a sub-field of $K$ (usually finitely generated). For $x, y \in K$, and $a \in P$, we can express the relation of *raising to power* $a \in P$ in terms of ex:

$$(y = x^a) \equiv \exists v \in K \; \mathrm{ex}\, v = x \; \& \; \mathrm{ex}\, av = y.$$

Obviously this is not a function but rather a 'multi-valued function'.

Of course, the theory of $K^P$ depends on Schanuel's conjecture, or rather its consequence for raising to powers in $P$. An instance of this conjecture, corresponding to the case when $P$ is generated by one 'generic' element, has been proven by A. Wilkie (unpublished).

There is an important advantage of considering $K^P$ rather than $K_{ex}$: it is easier to analyse; in particular, its first-order theory, unlike that of $K_{ex}$, does not interpret arithmetic. Moreover, the following is known.

**Theorem 6.5.10.** *The first-order theory of $K^P$ of characteristic zero is super-stable and has elimination of quantifiers to the level of Boolean combination of projective sets. Assuming Schanuel's conjecture for raising to powers $P \subseteq \mathbb{C}$, the theory $\mathbb{C}^P$ coincides with that of $K^P$.*

This is proved in a series of papers (Zilber, 2002a, 2003, 2004b). The analogue of Theorem B.2.1 is quite straightforward from the known facts.

Nevertheless, the problem of identifying the abstract $K^P$ as an analytic Zariski structure is open for all cases when $P \neq \mathbb{Q}$.

Other versions of two-sorted structures related to pseudo-exponentiation are the algebraically closed fields with 'real curves'. These are structures of the form $(K, +, \cdot, G)$, where $(K, +, \cdot)$ is an algebraically closed field and $G$ a unary predicate for a subset of distinguished points.

Several variations of the structure have been studied; we present these in slightly different versions than in the original papers of B. Poizat (2001).

(1) $G$ is a divisible subgroup of the multiplicative group $K^\times$ of $K$, obtained by Hrushovski's construction with the only condition that the pre-dimension of points in $G$ is half of the pre-dimension of $K$. Following B. Poizat (2001), $G$ is called the *green points subgroup*.

(2) $G$ is a subgroup of $K^\times$ of the form $G^0 \cdot \Gamma$, for $G^0$ divisible, $\Gamma$ an infinite cyclic subgroup. $G$ is obtained by the same modification of Hrushovski's construction as in Section B.2.3, with the condition that the pre-dimension of points in $G$ is half of the pre-dimension of $K$. We call this $G$ the *emerald points subgroup*.

**Theorem 6.5.11 (essentially B. Poizat).** *The first-order theories of the fields with green and emerald points are super-stable and have elimination of quantifiers to the level of Boolean combination of projective sets.*

*The model-theoretic dimension of the definable subgroup $G$ is*

$$U\text{-rank}(G) = \omega, \qquad U\text{-rank}(K) = \omega \cdot 2$$

*in both structures.*

In a previous work (Zilber, 2004a), it is shown that by assuming the Schanuel conjecture we can alternatively obtain the field with green points by interpreting $K$ as the complex numbers and $G$ as $G^0 \cdot \Gamma$, where $G^0$ is the 'spiral' $\exp(1 + i)\mathbb{R}$ $:= \{\exp t : t \in (1 + i)\mathbb{R}\}$ on the complex plane $\mathbb{C}$ and $\Gamma = \exp a\mathbb{Q}$, for some $a \in \mathbb{C} \setminus (1 + i)\mathbb{R}$.

The case of emerald points has a similar complex-real representation (forthcoming paper by J. D. Caycedo Casallas and the author), with the same $G^0$ but $\Gamma = a\mathbb{Z}$.

The 'analytic Zariski status' of the both examples at present is not quite clear even conjecturally but some nice properties of these as topological structures have been established. The case of emerald points is especially interesting as we can connect this structure to one of the central objects of non-commutative geometry, the quantum torus $T_\theta^2$ which may be defined graphically as the quotient-space $\mathbb{R}/L$, where $L$ is the subgroup $\mathbb{Z} + \theta\mathbb{Z}$ of $\mathbb{R}$. Indeed, for $a = 2\pi i \theta$, the following is definable in the structure with emerald points:

$$\mathbb{C}^\times / G \cong \mathbb{C}/((1 + i)\mathbb{R} + 2\pi i\mathbb{Z} + 2\pi i\theta\mathbb{Z}) \cong \mathbb{R}/(2\pi i\mathbb{Z} + 2\pi i\theta\mathbb{Z}).$$

If we also combine the structure of emerald points with that of raising to powers, we get a quantum torus 'with real multiplication'. Here 'real multiplication' is an endomorphism of the quantum torus in analogy with complex multiplication on an elliptic curve. One can check that analogously to the elliptic case $\mathbb{C}/((1 + i)\mathbb{R} + 2\pi i\mathbb{Z} + 2\pi i\theta\mathbb{Z})$ has an endomorphism induced by the multiplication $x \mapsto rx$ on $\mathbb{C}$, $r \notin \mathbb{Z}$, iff $\theta$ is a real quadratic irrational.

# Appendix A

## Basic model theory

## A.1 Languages and structures

The crucial feature of the model-theoretic approach to mathematics is the attention paid to the formalism with which one considers particular mathematical structures. We start by reviewing standard terminology and notation.

### Language: alphabet, terms, and formulas

The **alphabet of a language** $L$ consists of, by definition, the following symbols:

(i) relation symbols $P_i$, $(i \in I)$ and constant symbols $c_k$, $(k \in K)$ with some index sets $I, K$. Further, a positive integer $\rho_i$ is assigned to each $i \in I$ called the **arity** of the relation symbol $P_i$.

   The symbols in (i) are called **non-logical** symbols and also **primitives**, and their choice determines $L$. In addition, any language has the following symbols:

(ii) $\doteq$, the equality symbol;

(iii) $v_1, \ldots, v_n, \ldots$, the variables;

(iv) $\wedge, \vee, \neg$, the connectives;

(v) $\forall$ and $\exists$, the quantifiers; and

(vi) $(,),$ , parentheses and comma.

**Remark A.1.1.** Usually an alphabet would also contain function (operation) symbols, but this can be replaced by a relation symbol. To say $f(\bar{x}) = y$, we may use the expression $F(\bar{x}, y)$, where $F$ is a relation symbol of a corresponding arity. A language which does not use function symbols, like ours, is called a **relational language**.

   Words of the alphabet of $L$ constructed in a specific way are called $L$-formulas:

163

**Atomic $L$-formulas** are the words of the form $P(\tau_1, \ldots, \tau_\rho)$ for $P$, a relation $L$-symbol (including $\doteq$) of arity $\rho$, and $\tau_1, \ldots, \tau_\rho$ are variables or constant symbols.

We sometimes refer to an atomic formula $\varphi$ of the form $P(\tau_1, \ldots, \tau_\rho)$ as $\varphi(v_{i_1}, \ldots, v_{i_n})$ to mark the fact that all the variables occurring among $\tau_1, \ldots, \tau_\rho$ are in $v_{i_1}, \ldots, v_{i_n}$.

An $L$**-formula** is defined by the following recursive definition:

 (i)  any atomic $L$-formula is an $L$-formula;
 (ii)  if $\varphi$ is an $L$-formula, then so is $\neg\varphi$;
 (iii)  if $\varphi$, $\psi$ are $L$-formulas, then so is $(\varphi \wedge \psi)$;
 (iv)  if $\varphi$ is an $L$-formula, then so is $\exists v \varphi$ for any variable $v$; and
 (v)  nothing else is an $L$-formula.

We define the **complexity of an $L$-formula** $\varphi$ to be just the number of occurrences of $\wedge$, $\neg$, and $\exists$ in $\varphi$. It is obvious from the definition that an atomic formula is of complexity 0 and that any formula of complexity $l > 0$ is obtained by an application of (ii), (iii), or (iv) to formulas of less complexity.

For an atomic formula $\varphi(v_{i_1}, \ldots, v_{i_n})$, the distinguished variables are said to be **free** in $\varphi$. The variables which are free in $\varphi$ and $\psi$ in (ii) and (iii) are, by definition, also free in $\neg\varphi$ and $(\varphi \wedge \psi)$. The variable $v$ in (iv) is called **bounded** in $\exists v \varphi$, and the list of free variables for this formula is given by the free variables of $\varphi$ except $v$.

An $L$-formula with no free variables is called also an $L$**-sentence**.

We define a language $L$ to be the set of all $L$-formulas, thus $|L|$ is the cardinality of the set.

To give a meaning or **interpretation** of symbols of a language $L$, one introduces a notion of an $L$**-structure**. An $L$-structure $\mathcal{A}$ consists of

 (i)  a non-empty set $A$, called a **domain** or **universe** of the $L$-structure, and
 (ii)  an assignment of an element $c^{\mathcal{A}} \in A$ to any constant symbol $c$ of $L$.

Thus an $L$-structure is an object of the form

$$\mathcal{A} = \left\langle A; \{P_i^{\mathcal{A}}\}_{i \in I}; \{c_k^{\mathcal{A}}\}_{k \in K} \right\rangle.$$

$\{P_i^{\mathcal{A}}\}_{i \in I}$ and $\{c_k^{\mathcal{A}}\}_{k \in K}$ are called the **interpretations** of the predicate and constant symbols, respectively.

We write $A = \mathrm{Dom}(\mathcal{A})$.

If $\mathcal{A}$ and $\mathcal{B}$ are both $L$-structures, we say that $\mathcal{A}$ is **isomorphic** to $\mathcal{B}$, written $\mathcal{A} \cong \mathcal{B}$, if there is a bijection $\pi : \mathrm{Dom}(\mathcal{A}) \to \mathrm{Dom}(\mathcal{B})$ which **preserves** corresponding relation and constant symbols (i.e. for any $i \in I$ and $k \in K$):

(i) $\bar{a} \in P_i^{\mathcal{A}}$ iff $\pi(\bar{a}) \in P_i^{\mathcal{B}}$, and
(ii) $\pi(c_k^{\mathcal{A}}) = c_k^{\mathcal{B}}$.

The map $\pi$ is then called an **isomorphism**. If $\pi$ is only assumed to be injective but still satisfies (i) and (ii), then it is called an **embedding** and can be written as $\pi : \mathcal{A} \to \mathcal{B}$ or $\mathcal{A} \subseteq_\pi \mathcal{B}$.

**Assigning truth values to *L*-formulas in an *L*-structure**

Suppose $\mathcal{A}$ is an $L$-structure with domain $A$, $\varphi(v_1, \dots, v_n)$ is an $L$-formula with free variables $v_1, \dots, v_n$, and $\bar{a} = \langle a_1, \dots, a_n \rangle \in A^n$. Given these data, we assign a truth value **true**, written $\mathcal{A} \models \varphi(\bar{a})$, or **false**, $\mathcal{A} \nvDash \varphi(\bar{a})$, by the following rules:

(i) $\mathcal{A} \models P(a_1, \dots, a_n)$    iff    $\langle a_1, \dots, a_n \rangle \in P_i^{\mathcal{A}}$;
(ii) $\mathcal{A} \models \varphi_1(\bar{a}) \wedge \varphi_2(\bar{a})$    iff    $\mathcal{A} \models \varphi_1(\bar{a})$ and $\mathcal{A} \models \varphi_2(\bar{a})$;
(iii) $\mathcal{A} \models \neg\varphi(\bar{a})$    iff    $\mathcal{A} \nvDash \varphi(\bar{a})$;
(iv) $\mathcal{A} \models \exists v_n \varphi(a_1, \dots, a_{n-1}, v_n)$    iff    there is an $a_n \in A$ such that $\mathcal{A} \models \varphi(a_1, \dots, a_n)$; and
(v) and (vi) for $\vee$ and $\forall$ analogous to (ii) and (iv).

Given an $L$-structure $\mathcal{A}$ and an $L$-formula $\varphi(v_1, \dots, v_n)$, we can define the set

$$\varphi(\mathcal{A}) = \{\bar{a} \in A^n : \mathcal{A} \models \varphi(\bar{a})\}.$$

Sets of this form are called **definable**.

Because any subset of $A^n$ can be viewed as an $n$-ary relation, $\varphi(\bar{v})$ determines also an ***L*-definable relation**. If some $\varphi(\mathcal{A})$ coincides with a graph of a function $f : A^{n-1} \to A$, we say then that $f$ is an ***L*-definable function**.

**Exercise A.1.2.**

(i) *An embedding $\pi : \mathcal{A} \to \mathcal{B}$ of L-structures preserves atomic L-formulas (i.e. for any atomic $\varphi(v_1, \dots, v_n)$ for any $\bar{a} \in A^n$):*

     (∗)    $\mathcal{A} \models \varphi(\bar{a})$ *iff* $\mathcal{B} \models \varphi(\pi(\bar{a}))$.

(ii) *An isomorphism $\pi : \mathcal{A} \to \mathcal{B}$ between L-structures preserves any L-formula $\varphi(v_1, \dots, v_n)$ $(n \geq 0)$ (i.e. for any $\bar{a} \in A^n$):*

     (∗)    $\mathcal{A} \models \varphi(\bar{a})$ *iff* $\mathcal{B} \models \varphi(\pi(\bar{a}))$.

**Corollary A.1.3.** *For definable subsets (relations),*

$$\pi(\varphi(\mathcal{A})) = \varphi(\mathcal{B});$$

*in particular, when* $\pi : \mathcal{A} \to \mathcal{A}$ *is an automorphism,*

$$\pi(\varphi(\mathcal{A})) = \varphi(\mathcal{A}).$$

The latter is very useful in checking non-definability of some subsets or relations.

**Exercise A.1.4.** *The multiplication is not definable in* $\langle \mathbb{R}, + \rangle$.

### Agreement about notations

The proposition about the properties of isomorphic structures says that there is no harm in identifying elements of $\mathcal{A}$ with its images under an isomorphism. Correspondingly, when speaking about embedding $\pi : \mathcal{A} \to \mathcal{B}$, we identify $A = \mathrm{Dom}\,\mathcal{A}$ with its image $\pi(A) \subseteq B = \mathrm{Dom}\,\mathcal{B}$ element-wise. So, by default, $\mathcal{A} \subseteq \mathcal{B}$ assumes $A \subseteq B$.

Given two $L$-structures $\mathcal{A}$ and $\mathcal{B}$, we say that $\mathcal{A}$ is **elementarily equivalent to** $\mathcal{B}$, written $\mathcal{A} \equiv \mathcal{B}$, if for any $L$-sentence $\varphi$

$$\mathcal{A} \vDash \varphi \text{ iff } \mathcal{B} \vDash \varphi.$$

## A.2 Compactness theorem

Let $\Sigma$ be a set of $L$-sentences. We write $\mathcal{A} \vDash \Sigma$ if, for any $\sigma \in \Sigma$, $\mathcal{A} \vDash \sigma$.

An $L$-sentence $\sigma$ is said to be a **logical consequence** of a finite $\Sigma$, written $\Sigma \vDash \sigma$, if $\mathcal{A} \vDash \Sigma$ implies $\mathcal{A} \vDash \sigma$ for every $L$-structure $\mathcal{A}$. For $\Sigma$ infinite, $\Sigma \vDash \sigma$ means that there is a finite $\Sigma^0 \subset \Sigma$ such that $\Sigma^0 \vDash \sigma$.

The symbol $\sigma$ is called **logically valid**, written $\vDash \sigma$, if $\mathcal{A} \vDash \sigma$ for every $L$-structure $\mathcal{A}$.

A set $\Sigma$ of $L$-sentences is said to be **satisfiable** if there is an $L$-structure $\mathcal{A}$ such that $\mathcal{A} \vDash \Sigma$. $\mathcal{A}$ is then called **a model** of $\Sigma$.

$\Sigma$ is said to be **finitely satisfiable (f.s.)** if any finite subset of $\Sigma$ is satisfiable.

$\Sigma$ is said to be **complete** if, for any $L$-sentence $\sigma$, $\sigma \in \Sigma$ or $\neg\sigma \in \Sigma$.

We would need to sometimes expand or reduce our language.

Let $L$ be a language with non-logical symbols $\{P_i\}_{i \in I} \cup \{c_k\}_{k \in K}$ and $L' \subseteq L$ with non-logical symbols $\{P_i\}_{i \in I'} \cup \{c_k\}_{k \in K'}$ ($I' \subseteq I$, $K' \subseteq K$). Let

$$\mathcal{A} = \langle A; \{P_i^{\mathcal{A}}\}_{i \in I}; \{c_k^{\mathcal{A}}\}_{k \in K} \rangle$$

and

$$\mathcal{A}' = \langle A; \{P_i^{\mathcal{A}}\}_{i \in I'}; \{c_k^{\mathcal{A}}\}_{k \in K'} \rangle.$$

Under these conditions, we call $\mathcal{A}'$ the $L'$-**reduct of** $\mathcal{A}$ and, correspondingly, $\mathcal{A}$ is an $L$-**expansion of** $\mathcal{A}'$.

**Remark A.2.1.** Obviously, under the notations for an $L'$-formula $\varphi(v_1, \ldots, v_n)$ and $a_1, \ldots, a_n \in A$,

$$\mathcal{A}' \vDash \varphi(a_1, \ldots, a_n) \text{ iff } \mathcal{A} \vDash \varphi(a_1, \ldots, a_n).$$

**Theorem A.2.2 (compactness theorem).** *Any finitely satisfiable set of L-sentences $\Sigma$ is satisfiable. Moreover, $\Sigma$ has a model of cardinality less or equal to $|L|$, the cardinality of the language.*

This is usually proven by the method called Henkin's construction, producing a model of $\Sigma$, elements of which are constructed from constant symbols of an expansion $L^\sharp$ of $L$. The relation between these elements is described by atomic $L^\sharp$-sentences which can be derived from $\Sigma^\sharp$, a completion of $\Sigma$. Of course, $\Sigma^\sharp$ is not determined by $\Sigma$ but is found by applying the Zorn lemma. So, in general, this is not an effective construction.

An embedding of $L$ structures $\pi : \mathcal{A} \to \mathcal{B}$ is called **elementary** if $\pi$ preserves any $L$-formula $\varphi(v_1, \ldots, v_n)$ (i.e. for any $a_1, \ldots, a_n \in \text{Dom}\,\mathcal{A}$)

$$\mathcal{A} \vDash \varphi(a_1, \ldots, a_n) \qquad \text{iff} \qquad \mathcal{B} \vDash \varphi(\pi(a_1), \ldots, \pi(a_n)). \tag{A.1}$$

We write the fact of elementary embedding as

$$\mathcal{A} \preccurlyeq \mathcal{B}.$$

It is often convenient to consider a partial elementary embedding $\pi$ that is defined on $D \subseteq \text{Dom}\,\mathcal{A}$ only. In this case, the definition requires that (A.1) holds for $a_1, \ldots, a_n \in D$ only.

In this case, we say $\pi$ is an **elementary monomorphism** $D \to \mathcal{B}$.

We usually identify $A = \text{Dom}\,\mathcal{A}$ with the subset $\pi(A)$ of $B = \text{Dom}\,\mathcal{B}$. Then $\pi(a) = a$ for all $a \in A$, and so $\mathcal{A} \preccurlyeq \mathcal{B}$ usually means

$$\mathcal{A} \vDash \varphi(a_1, \ldots, a_n) \qquad \text{iff} \qquad \mathcal{B} \vDash \varphi(a_1, \ldots, a_n).$$

For an $L$-structure $\mathcal{A}$, let $L_A = L \cup \{c_a : a \in A\}$ be the expansion of the language and $\mathcal{A}^+$ be the natural expansion of $\mathcal{A}$ to $L_A$, assigning to $c_a$ the element $a$. The **diagram of** $\mathcal{A}$ is

Diag$(\mathcal{A}) = \{\sigma : \sigma$ an atomic $L_A$-sentence or negation of

an atomic $L_A$-sentence, such that $\mathcal{A}^+ \vDash \sigma\}$.

The **complete diagram of** $\mathcal{A}$ is defined as

$$\text{CDiag}(\mathcal{A}) = \{\sigma : \sigma \ L_A\text{-sentence such that} \quad \mathcal{A}^+ \vDash \sigma\}.$$

**Theorem A.2.3 (method of diagrams).** *For an L structure $\mathcal{B}$,*

(i) *there is an expansion $\mathcal{B}^+$ to the language $L_A$ such that $\mathcal{B}^+ \models \mathrm{Diag}(\mathcal{A})$ iff $\mathcal{A} \subseteq \mathcal{B}$, and*

(ii) *there is an expansion $\mathcal{B}^+$ to the language $L_A$ such that $\mathcal{B}^+ \models \mathrm{CDiag}(\mathcal{A})$ iff $\mathcal{A} \preccurlyeq \mathcal{B}$.*

*Proof.* Indeed, by definitions and Exercise A.1.2, $a \to c_a^{\mathcal{B}^+}$ is an embedding iff $\mathcal{B}^+ \models \mathrm{Diag}(\mathcal{A})$.

The elementary embedding case is straightforward by definition. □

**Corollary A.2.4.** *Given an L-structure $\mathcal{A}$ and a set of L-sentences $T$,*

(i) *the set $T \cup \mathrm{Diag}(\mathcal{A})$ is finitely satisfiable iff there is a model $\mathcal{B}$ of $T$ such that $\mathcal{A} \subseteq \mathcal{B}$, and*

(ii) *the set $T \cup \mathrm{CDiag}(\mathcal{A})$ is finitely satisfiable iff there is a model $\mathcal{B}$ of $T$ such that $\mathcal{A} \preccurlyeq \mathcal{B}$.*

**Theorem A.2.5 (upward Lowenheim–Skolem theorem).** *For any infinite L-structure $\mathcal{A}$ and a cardinal $\kappa \geq \max\{|L|, \|\mathcal{A}\|\}$, there is an L-structure $\mathcal{B}$ of cardinality $\kappa$ such that $\mathcal{A} \preccurlyeq \mathcal{B}$.*

*Proof.* Let $M$ be a set of cardinality $\kappa$. Consider an extension $L_{A,M}$ of language $L$ obtained by adding to $L_A$ constant symbols $c_i$ for each $i \in M$. Consider now the set of $L_{A,M}$-sentences

$$\Sigma = \mathrm{CDiag}(\mathcal{A}) \cup \{\neg c_i \doteq c_j : i \neq j \in M\}.$$

It is easy to see that $\Sigma$ is finitely satisfiable.

It follows from the compactness theorem that $\Sigma$ has a model of cardinality $|L_{A,M}|$, which is equal to $\kappa$. Let $\mathcal{B}^*$ be such a model. The $L$-reduct $\mathcal{B}$ of $\mathcal{B}^*$, by the method of diagrams, satisfies the requirement of the theorem. □

**Corollary A.2.6.** *Let $\Sigma$ be a set of L-sentences which has an infinite model. Then for any cardinal $\kappa \geq |L|$, there is a model of $\Sigma$ of cardinality $\kappa$.*

**Theorem A.2.7 (Tarski–Vaught test).** *Suppose $\mathcal{A} \subseteq \mathcal{B}$ are L-structures with domains $A \subseteq B$. Then $\mathcal{A} \preccurlyeq \mathcal{B}$ iff the following condition holds: for all L-formulas $\varphi(v_1, \ldots, v_n)$ and all $a_1, \ldots, a_{n-1} \in A$, $b \in B$ such that $\mathcal{B} \models \varphi(a_1, \ldots, a_{n-1}, b)$, there is $a \in A$ with $\mathcal{B} \models \varphi(a_1, \ldots, a_{n-1}, a)$.*

**Definition A.2.8.** Let

$$\mathcal{A}_0 \subseteq \mathcal{A}_1 \subseteq \cdots \subseteq \mathcal{A}_i \subseteq \cdots \tag{A.2}$$

be a sequence of $L$-structures, $i \in \mathbb{N}$, forming a chain with respect to embeddings. Denote $\mathcal{A}^* = \bigcup_n \mathcal{A}_n$ the $L$-structure with

the domain $A^* = \bigcup_n A_n$,
predicates $P^{\mathcal{A}^*} = \bigcup_n P^{\mathcal{A}_n}$, for each predicate symbol $P$ of $L$, and
$c^{\mathcal{A}^*} = c^{\mathcal{A}_0}$, for each constant symbol from $L$.

By definition $\mathcal{A}_n \subseteq \mathcal{A}^*$, for each $n$.

**Exercise A.2.9.** *Use the induction on complexity of formulas and the Tarski–Vaught test to prove that if in Equation (A.2) for each $n$, $\mathcal{A}_n \preccurlyeq \mathcal{A}_{n+1}$ that is, the chain is elementary, then $\mathcal{A}_n \preccurlyeq \mathcal{A}^*$, for each $n$.*

A class $C$ of $L$-structures is called **axiomatizable** if there is a set $\Sigma$ of $L$-sentences such that

$$\mathcal{A} \in C \qquad \text{iff} \qquad \mathcal{A} \models \Sigma.$$

We also write equivalently

$$C = \text{Mod}\,(\Sigma).$$

$\Sigma$ is then called **a set of axioms for $C$.**

The **theory of $C$**, Th($C$), is the set of $L$-sentences which hold in any structure of the class $C$. Obviously, $\Sigma \subseteq \text{Th}(C)$. One can also say that $\Sigma$ is a **set of axioms** for Th($C$).

A formula of the form $\exists v_1 \ldots \exists v_m \theta$, where $\theta$ is a quantifier-free formula, is called an **existential formula** (or an $\exists$-formula). The negation of an existential formula is called a **universal** ($\forall$-formula) formula.

**Exercise A.2.10.** *Let $\phi_1, \ldots, \phi_n$ be existential formulas. Prove that*

*(i) $(\phi_1 \vee \cdots \vee \phi_n)$ and $(\phi_1 \wedge \cdots \wedge \phi_n)$ are logically equivalent to existential formulas; and*
*(ii) $(\neg\phi_1 \wedge \cdots \wedge \neg\phi_n)$ and $(\neg\phi_1 \vee \cdots \vee \neg\phi_n)$ are logically equivalent to universal formulas.*

**Exercise A.2.11.** *Suppose $\mathcal{A} \subseteq \mathcal{B}$ and $a_1, \ldots, a_n \in A$.*

*(i) If $\mathcal{A} \models \varphi(a_1, \ldots, a_n)$, for an existential formula $\varphi(v_1, \ldots, v_n)$, then $\mathcal{B} \models \varphi(a_1, \ldots, a_n)$.*
*(ii) If $\mathcal{B} \models \psi(a_1, \ldots, a_n)$, for a universal formula $\psi(v_1, \ldots, v_n)$, then $\mathcal{A} \models \psi(a_1, \ldots, a_n)$.*

An axiomatizable class $C$ is said to be $\forall$**-axiomatizable** ($\exists$-axiomatizable) if $\Sigma$ can be chosen to consist of universal (existential) sentences only.

An $L$-formula is said to be **positive** if it is equivalent to a formula which does not contain the negation $\neg$. A **positively axiomatizable class** is a class axiomatizable by a set of positive axioms. Given a set of $L$-sentences $\Sigma$, we use $\Sigma_+$ to denote the subset of $\Sigma$ consisting of positive sentences.

**Exercise A.2.12.** *Let $C$ be a positively axiomatizable class. Prove that if $\mathcal{A}$, $\mathcal{B}$ are $L$-structures, $\mathcal{A} \in C$, and there is a surjective homomorphism $h : \mathcal{A} \to \mathcal{B}$, then $\mathcal{B} \in C$. That is, $C$ is closed under homomorphisms.*

The following is one of the typical *preservation theorems* proved in the 1950s. The proof (Hodges 1993) is quite intricate but uses no more than the compactness theorem.

**Theorem A.2.13 (R. Lyndon).** *Let $C$ be an axiomatizable class. Then $C$ is positively axiomatizable iff it is closed under homomorphisms.*

## A.3 Existentially closed structures

**Definition A.3.1.** Let $C$ be a class of $L$-structures and $\mathcal{A} \in C$. We say that $\mathcal{A}$ is **existentially closed** in $C$ if for every quantifier-free $L$-formula $\psi(\bar{v}, \bar{w})$, for any $\bar{a}$ in $\mathcal{A}$ and any $\mathcal{B} \supseteq \mathcal{A}$, $\mathcal{B} \in C$,

$$\mathcal{B} \vDash \exists \bar{v}\psi(\bar{v}, \bar{a}) \Rightarrow \mathcal{A} \vDash \exists \bar{v}\psi(\bar{v}, \bar{a}).$$

**Exercise A.3.2.** *Algebraically closed fields are exactly the existentially closed objects in the class of all fields.*

**Theorem A.3.3.** *Let $T$ be a theory such that every model $\mathcal{A}$ of $T$ is existentially closed in the class of models of $T$. Then*

$$\mathcal{A} \subseteq \mathcal{B} \Rightarrow \mathcal{A} \preccurlyeq \mathcal{B}$$

*for any two models $\mathcal{A}$ and $\mathcal{B}$ of $T$.*

*Proof.* Given $\mathcal{A} \subseteq \mathcal{B}$, models of $T$, first note that there is $\mathcal{A}' \succcurlyeq \mathcal{A}$ such that $\mathcal{A} \subseteq \mathcal{B} \subseteq \mathcal{A}'$. Indeed, by Corollary A.2.4, it suffices to show that CDiag($\mathcal{A}$) $\cup$ Diag($\mathcal{B}$) is finitely satisfiable. This amounts to checking that for any quantifier-free $\psi(\bar{v}, \bar{a})$ with $\bar{a}$ in $\mathcal{A}$, the set CDiag($\mathcal{A}$) $\cup$ $\{\exists \bar{v}\psi(\bar{v}, \bar{a})\}$ is satisfiable, but $\mathcal{A}^+$ is the model of the set by the assumption of the theorem.

Now we can go on repeating this construction to produce the chain

$$\mathcal{A}_0 \subseteq \mathcal{B}_0 \subseteq \mathcal{A}_1 \subseteq \mathcal{B}_1 \subseteq \cdots$$

such that

$$\mathcal{A} = \mathcal{A}_0 \preccurlyeq \mathcal{A}_1 \preccurlyeq \mathcal{A}_2 \preccurlyeq \cdots$$

and

$$\mathcal{B} = \mathcal{B}_0 \preccurlyeq \mathcal{B}_1 \preccurlyeq \mathcal{B}_2 \preccurlyeq \cdots$$

Consider

$$\mathcal{A}^* = \bigcup_i \mathcal{A}_i = \bigcup_i \mathcal{B}_i = \mathcal{B}^*.$$

Then, by the Exercise A.2.9, $\mathcal{A} \prec \mathcal{A}^* \succ \mathcal{B}$ which implies $\mathcal{A} \preccurlyeq \mathcal{B}$. $\qquad\square$

**Definition A.3.4.** A theory $T$ satisfying the conclusion of Theorem A.3.3 is said to be **model complete**.

Note that if $T$ is model complete, then the assumption of Theorem A.3.3 is satisfied; that is, any model is existentially closed.

**Theorem A.3.5.** *Any formula $\varphi(\bar{v})$ in a model-complete theory, $T$, is equivalent to an $\exists$-formula $\psi(\bar{v})$.*

Note that model completeness of a theory is of geometric significance similar to quantifier elimination. It says that in such a theory one does not have to deal with sets more complex than **projective** ones, that is, definable by $\exists$-formulas. A projection of an arbitrary Boolean combination of such sets is just another projective set.

The condition on the existence of existentially closed models in a class $C$ is very simple. Call a class $C$ **inductive** if the union of any ascending (transfinite) chain of structures of $C$ again belongs to $C$.

**Exercise A.3.6.** *Any $\mathcal{A}$ of an inductive class $C$ can be embedded into an existentially closed $\mathcal{B} \in C$.*

More recently, it has been realised (by S. Shelah and E. Hrushovski first of all) that it is useful to consider existential closedness in classes $C$ with a restricted notion of embeddings, written as $\mathcal{A} \leq \mathcal{B}$, of structures. The statement A.3.6 still holds for classes $(C, \leq)$. The sub-classes of existentially closed structures might be quite complicated but under some natural conditions have good properties, in particular the property that definable sets are just Boolean combinations of projective ones. See Section B.2.2 for a further discussion of the notion.

## A.4 Complete and categorical theories

We continue our discussion of axiomatizable classes, but now our interest is mainly in those which are axiomatised by a complete set of axioms.

**Definition A.4.1.** A theory $T$ is said to be **categorical in power (cardinality)** $\kappa$ ($\kappa$-categorical) if there is a model $\mathcal{A}$ of $T$ of cardinality $\kappa$ and any model of $T$ of this cardinality is isomorphic to $\mathcal{A}$.

Note that the **absolute categoricity** of $T$, requiring that there is just a unique, up-to-isomorphism, model of $T$, is not very useful in the first-order context. Indeed, by the Löwenheim–Skolem theorem, the unique model can be only finite.

**Theorem A.4.2 (R. Vaught).** *Let $\kappa \geq |L|$ and $T$ be a $\kappa$-categorical L-theory without finite models. Then $T$ is complete.*

*Proof.* Let $\sigma$ be an $L$-sentence and $\mathcal{A}$ the unique, up-to-isomorphism, model of $T$ of cardinality $\kappa$. Either $\sigma$ or $\neg\sigma$ holds in $\mathcal{A}$; let it be $\sigma$. Then $T \cup \{\neg\sigma\}$ does not have a model of cardinality $\kappa$, which by the Löwenheim–Skolem theorems means $T \cup \{\neg\sigma\}$ does not have an infinite model, which by our assumption means it is not satisfiable. It follows that $T \vDash \sigma$. $\square$

So, categoricity in powers is a stronger form of completeness.

**Example A.4.3.** *Let $K$ be a field and $L_K$ be the language with alphabet $\{+, \lambda_k, 0\}_{k \in K}$ where $+$ is a symbol of a binary function, $\lambda_k$ are symbols of unary functions, and 0 is constant symbol. Define the* **theory of $K$-vector spaces** *by the following well-known axioms:*

$\forall v_1 \forall v_2 \forall v_3 \ (v_1 + v_2) + v_3 \simeq v_1 + (v_2 + v_3);$
$\forall v_1 \forall v_2 \ v_1 + v_2 \simeq v_2 + v_1;$
$\forall v \ v + 0 \simeq v;$
$\forall v_1 \exists v_2 \ v_1 + v_2 \simeq 0;$
$\forall v_1 \forall v_2 \ \lambda_k(v_1 + v_2) \simeq \lambda_k(v_1) + \lambda_k(v_2),$ *an axiom for each $k \in K$;*
$\forall v \ \lambda_1(v) \simeq v;$
$\forall v \ \lambda_0(v) \simeq 0;$
$\forall v \ \lambda_{k_1}(\lambda_{k_2}(v)) \simeq \lambda_{k_1 \cdot k_2}(v),$ *an axiom for each $k_1, k_2 \in K$; and*
$\forall v \ \lambda_{k_1}(v) + \lambda_{k_2}(v) \simeq \lambda_{k_1 + k_2}(v),$ *an axiom for each $k_1, k_2 \in K$.*

Let $\mathcal{A}$ be a model of the theory of $K$-vector spaces (that is a $K$-vector space) of cardinality $\kappa > |L_K| = \max\{\aleph_0, \text{card } K\}$. Then the cardinality of $\mathcal{A}$ is equal to the dimension of the vector space. It follows that if $\mathcal{B}$ is another model of

$V_K$ of the same cardinality, then $\mathcal{A} \cong \mathcal{B}$. Thus, we have checked the validity of the following statement.

**Theorem A.4.4.** *The theory of K-vector spaces is $\kappa$-categorical for any $\kappa >$ card K.*

**Example A.4.5.** *Let L be the language with one binary symbol $<$ and DLO be the theory of dense linear order with no end elements:*

$\forall v_1 \forall v_2 \ (v_1 < v_2 \rightarrow \neg \ v_2 < v_1)$;
$\forall v_1 \forall v_2 \ (v_1 < v_2 \vee v_1 \doteq v_2 \vee v_2 < v_1)$;
$\forall v_1 \forall v_2 \forall v_3 \ (v_1 < v_2 \wedge v_2 < v_3) \rightarrow v_1 < v_3$;
$\forall v_1 \forall v_2 \ (v_1 < v_2 \rightarrow \exists v_3 \ (v_1 < v_3 \wedge v_3 < v_2))$; and
$\forall v_1 \exists v_2 \exists v_3 \ v_1 < v_2 \wedge v_3 < v_1$.

**Theorem A.4.6 (G. Cantor).** *Any two countable models of DLO are isomorphic. In other words, DLO is $\aleph_0$-categorical.*

To prove that any two countable models of DLO are isomorphic, we enumerate the two ordered sets and then apply the famous *back-and-forth construction* of a bijection preserving the orders. More details on the method of proof are in Remark A.4.21.

**Exercise A.4.7.** *Show that DLO is not $\kappa$-categorical for any $\kappa > \aleph_0$.*

**Example A.4.8.** *The* **theory of algebraically closed fields of characteristic** *$p$, $\mathrm{ACF}_p$, for $p$ a prime number or $0$, is given by the following axioms in the language of fields $L_{\text{fields}}$ with binary operations $+$ and $\cdot$ and constant symbols $0$ and $1$.*

1. Axioms of fields:

$\forall v_1 \forall v_2 \forall v_3$
$$(v_1 + v_2) + v_3 \doteq v_1 + (v_2 + v_3)$$
$$(v_1 \cdot v_2) \cdot v_3 \doteq v_1 \cdot (v_2 \cdot v_3)$$
$$v_1 + v_2 \doteq v_2 + v_1$$
$$v_1 \cdot v_2 \doteq v_2 \cdot v_1$$
$$(v_1 + v_2) \cdot v_3 \doteq v_1 \cdot v_3 + v_2 \cdot v_3$$
$$v_1 + 0 \doteq v_1$$
$$v_1 \cdot 1 \doteq v_1.$$
$\forall v_1 \exists v_2 \ v_1 + v_2 \doteq 0$
$\forall v_1 (\neg v_1 \doteq 0 \rightarrow \exists v_2 \ v_1 \cdot v_2 \doteq 1)$.

2. Axiom stating that the field is of characteristic $p > 0$:

$$\underbrace{1 + \cdots + 1}_{p} \simeq 0.$$

To state that the field is of characteristic 0, one has to write down the infinite list of axioms, one for each positive integer $n$:

$$\neg \underbrace{1 + \cdots + 1}_{n} \simeq 0.$$

3. Solvability of polynomial equation axioms, one for each positive integer $n$:

$$\forall v_1 \ldots \forall v_n \exists v \ v^n + v_1 \cdot v^{n-1} + \cdots + v_i \cdot v^i + \cdots + v_n \simeq 0.$$

**Remark A.4.9.** It is easy to see that any quantifier-free formula in the language $L_{\text{fields}}$ can be replaced by an equivalent quantifier-free formula in the language of $L_{\text{Zar}}$ (see Example 1.2.3) and conversely. Moreover, positive formulas correspond to positive formulas, so the two languages are essentially equivalent.

Recall that a **transcendence basis of a field** $K$ is a maximal algebraically independent subset of $K$. **The transcendence degree of a field** $K$ is the cardinality of a basis of the field.

**Steinitz theorem.** *If $B_1$ is a basis of $K_1$ and $B_2$ is a basis of $K_2$, algebraically closed fields of the same characteristic, and $\pi : B_1 \to B_2$ is a bijection, then $\pi$ can be extended to an isomorphism between the fields.*

In other words, the isomorphism type of an algebraically closed field of a given characteristic is determined by its transcendence degree. Also, the transcendence degree of a field $K$ is equal to the cardinality of the field modulo $\aleph_0$. In other words, for uncountable fields $\text{tr.d.} K = \text{card } K$.

It follows that if $K_1$ and $K_2$ are two models of $\text{ACF}_p$ of an uncountable cardinality $\kappa$, then $K_1 \cong K_2$.

**Theorem A.4.10.** *$\text{ACF}_p$ is categorical in any uncountable power $\kappa$.*

(More detail on this is in a later sub-section). It is also useful to consider the following simple examples.

**Example A.4.11 (free $G$-module).** *Let $G$ be a group and $L_G$ be the language with unary function symbols $g$, each $g \in G$. The axioms of the theory $T_G$ say that*

  *(i) the functions $g$ corresponding to each function symbol are invertible;*

*(ii) the composition $g_1 g_2$ is equal to g iff the equality holds in the group; and*
*(iii) $g(x) = x$ for an element x iff $g = e$, the identity of the group.*

Any model of $T_G$ is the union of non-intersecting free orbits of G, so the theory is categorical in all infinite cardinalities $\kappa$ greater than card G.

Note that when G is the trivial group, $G = \{e\}$, $T_G$ is in fact the **trivial theory** in the trivial language containing only the equality.

### A.4.1 Types in complete theories

Fix a language L. Henceforth, T denotes a complete L-theory having an infinite model, say $\mathcal{A}$. By the Lowenheim–Skolem downward theorem, we may assume $\mathcal{A}$ is of cardinality equal to that of L. Also, by definition, $T = \text{Th}(\mathcal{A})$.

**Definition A.4.12.** An *n*-**type** p (in T) is a set of formulas with n free variables $\bar{v} = (v_1, \dots, v_n)$, such that

(i) for all $\varphi \in p$, $T \models \exists \bar{v} \varphi(\bar{v})$;
(ii) if $\varphi, \psi \in p$, then $(\varphi \wedge \psi) \in p$.
   Type p is called **complete** if the following is also satisfied:
(iii) for any $\varphi \in F_n$, either $\varphi \in p$ or $\neg \varphi \in p$.

Suppose $\bar{a} \in A^n$. Then we define **the L-type of $\bar{a}$ in $\mathcal{A}$.**

$$\text{tp}_{\mathcal{A}}(\bar{a}) = \{\varphi \in F_n : \mathcal{A} \models \varphi(\bar{a})\}.$$

Clearly, $\text{tp}_{\mathcal{A}}(\bar{a})$ is a complete *n*-type.

When $\mathcal{A} \subseteq \mathcal{B}$, then $\text{tp}_{\mathcal{A}}(a)$ and $\text{tp}_{\mathcal{B}}(a)$ may be different, but it follows immediately from definitions that

$$\mathcal{A} \preccurlyeq \mathcal{B} \text{ implies } \text{tp}_{\mathcal{A}}(a) = \text{tp}_{\mathcal{B}}(a).$$

We say that an *n*-type p is **realised** in $\mathcal{A}$ if there is $\bar{a} \in A^n$ such that $p \subseteq \text{tp}_{\mathcal{A}}(\bar{a})$. If there is no such $\bar{a}$ in $\mathcal{A}$, we say that p is **omitted** in $\mathcal{A}$.

**Exercise A.4.13.** *Given a set $P = \{p^\alpha : \alpha < \kappa\}$ of n-types p, a model $\mathcal{A}$ of T and a cardinal $\kappa \geq |\mathcal{A}|$, there is $\mathcal{B} \succcurlyeq \mathcal{A}$ of cardinality $\kappa$ such that all types from P are realised in $\mathcal{B}$.*

**Corollary A.4.14.** *For any n-type, there is $p' \supseteq p$ which is a complete n-type.*

Indeed, put $p' = \text{tp}_{\mathcal{B}}(\bar{a})$ for $\bar{a}$ in $\mathcal{B}$ realising p.

**Remark A.4.15.** If $\pi : \mathcal{A} \to \mathcal{B}$ is an isomorphism, $\bar{a} \in A^n$, $\bar{b} \in B^n$, and $\pi : \bar{a} \mapsto \bar{b}$, then $\text{tp}_A(\bar{a}) = \text{tp}_B(\bar{b})$.

The statement that a given theory $T$ in a language $L$ allows **quantifier elimination** means that for every $L$-formula $\psi(v_1, \ldots, v_n)$ with $n$ free variables, there is a quantifier-free $L$-formula $\varphi(v_1, \ldots, v_n)$ such that

$$T \models \psi \leftrightarrow \varphi.$$

Here, $\text{qftp}(a/A)$ stands for the quantifier-free type (consisting of quantifier-free formulas only) of $a$ over $A$.

In this sub-section, we are going to demonstrate a method of proving quantifier elimination. This method applies a basic algebraic analysis of models of a given theory and as a by-product also produces useful algebraic information.

**Theorem A.4.16 (A. Tarski).** *$\text{ACF}_p$ is complete and allows quantifier elimination in the language $L_{\text{Zar}}$.*

*Proof.* The completeness of $\text{ACF}_p$ follows from categoricity (Theorem A.4.10). It remains to prove quantifier elimination. We first prove the following statement, in fact closely following the standard proof of the Steinitz theorem.

**Lemma A.4.17.** *Let $K$ be an algebraically closed field and $k_0$ be its prime sub-field. For any $A \subseteq K$, any two $n$-tuples $\bar{b}$ and $\bar{c}$,   $\text{qftp}(\bar{b}/A) = \text{qftp}(\bar{c}/A)$ iff $\bar{b}$ is conjugated with $\bar{c}$ by an automorphism over $A$ iff $\text{tp}(\bar{b}/A) = \text{tp}(\bar{c}/A)$.*

*Proof.* First consider $n = 1$. Use $k_0(A)$ to denote the sub-field generated by $A$, and use $k_0(A)(x)$ to denote the field of rational functions over it. If $b$ is transcendental over $k_0(A)$, then so is $c$ and

$$k_0(Ab) \cong k_0(A)(x) \cong k_0(Ac) \text{ over } A.$$

If $f(x)$ is the minimal polynomial of $b$ over $k_0(A)$, then so is $f(x)$ with respect to $c$ and

$$k_0(Ab) \cong k_0(A)[x]/\{f(x)\} \cong k_0(Ac) \text{ over } A.$$

If  $\bar{b} = \langle b_1 \ldots, b_n \rangle$  and  $\text{qftp}(\bar{b}/A) = \text{qftp}(\bar{c}/A)$,  then  $\text{qftp}(b_1/A) = \text{qftp}(c_1/A)$. Thus, by induction there is an isomorphism

$$\alpha : k_0(Ab_1) \to k_0(Ac_1).$$

Let $\langle b_2' \ldots, b_n' \rangle$ be the image of $\langle b_2 \ldots, b_n \rangle$ under $\alpha$. Then

$$\text{qftp}(\langle c_1, b_2' \ldots, b_n' \rangle / A) = \text{qftp}(\bar{b}/A) = \text{qftp}(\bar{c}/A),$$

and hence

$$\langle b'_2 \ldots, b'_n \rangle \qquad \text{conjugated with} \qquad \langle c_2, \ldots, c_n \rangle \text{ over } Ac_1.$$

Finally $\bar{b}$ is conjugated with $\bar{c}$ over $A$. □

**End of Proof of Theorem A.4.16.** Let $\varphi(\bar{x})$ be any formula in the language $L_{\text{Zar}}$ and

$$\Phi(\bar{x}) = \{\psi(\bar{x}) \text{ qfree} : K \models \varphi(\bar{x}) \rightarrow \psi(\bar{x})\}.$$

If $\Phi \& \neg\varphi$ is consistent, then in the universal domain $^*F \succ F$ there is a realisation $\bar{b}$ of the type. Qftp($\bar{b}$) must be consistent with $\varphi$, for otherwise $\neg\xi(\bar{x})$ is in $\Phi$ for some $\xi \in$ qftp($\bar{b}$). Then there exists $\bar{c}$ realizing qftp($\bar{b}$)$\&\varphi$, which is a contradiction.

Thus, $\models \Phi \rightarrow \varphi$ and so $\Phi$ is equivalent to its finite part and $\Phi \equiv \varphi$. □

Quantifier elimination is a powerful tool and essentially the condition for a successful application of model theory in concrete fields of mathematics. It also has a distinctive geometric significance when it takes place in a natural language. In particular, the quantifier elimination for $\text{ACF}_p$ is a crucial tool in algebraic geometry – it effectively says that *the image of a quasi-projective variety under a morphism is quasi-projective* (Chevalley, Seidenberg, Tarski).

**Exercise A.4.18.** *Prove quantifier elimination theorems for*

* *the theory $V_K$ of vector spaces and*
* *the theory* DLO *of dense linear orders without endpoints.*

### A.4.2 Spaces of types and saturated models

Let $T$ be a complete theory of a countable language $L$.

We use $S_n(T)$ to denote the set of all complete $n$-types in $T$, the **(Stone) space of $n$-types of** $T$. There is a standard **Stone topology** on $S_n(T)$, the basis of open sets of which is given by sets of the form

$$U_\psi = \{p \in S_n(T) : \psi \in p\}$$

for each $L$-formula $\psi$ with $n$ free variables $v_1, \ldots, v_n$.

It is easy to see that the Stone space is compact in the topology.

**Definition A.4.19.** Given an infinite cardinal $\kappa$, a structure $\mathcal{A}$ is called $\kappa$-**saturated** if, for any cardinal $\lambda < \kappa$, for any expansion $\mathcal{A}_C$ of $\mathcal{A}$ by constant symbols $C = \{c_i : i \leq \lambda\}$, every 1-type in Th($\mathcal{A}_C$) is realised in $\mathcal{A}_C$.

We say just **saturated** instead of $\kappa$-saturated when $\kappa = \text{card } A$.

A model $\mathcal{A}$ of $T$ is called $\kappa$-**universal** if, for any model $\mathcal{B}$ of $T$ of cardinality not bigger than $\kappa$, there is an elementary embedding $\pi : \mathcal{B} \to \mathcal{A}$.

**Theorem A.4.20.** *Let $T$ be a complete theory.*

*(i) Any $\kappa$-saturated model of $T$ is $\kappa$-universal.*
*(ii) Any two saturated models of $T$ of the same cardinality are isomorphic.*
*(iii) For every $\kappa \geq \aleph_0$, there exists a $\kappa$-saturated model of $T$.*

*Proof.* For (i) and (ii), we use a standard inductive construction.

(i) Let $\mathcal{A}$ be a $\kappa$-saturated model and $\mathcal{B}$ a model of $T$ of cardinality $\lambda \leq \kappa$. This means that we can present the domain $B$ of $\mathcal{B}$ as

$$B = \{b_i : 0 \leq i \leq \lambda\}.$$

We construct by induction on $\alpha < \lambda$ the sequence $\{a_i \in A : 0 \leq i < \alpha\} \subseteq A$ with the property

$$\mathcal{A} \models \psi(a_{i_1}, \ldots, a_{i_n}) \qquad \text{iff} \qquad \mathcal{B} \models \psi(b_{i_1}, \ldots, b_{i_n})$$

for any formula $\psi$ in $n$ free variables and any $i_1, \ldots, i_n < \alpha$.

For $a_0$, we take any element which satisfies the type $\text{tp}_{\mathcal{B}}(b_0)$. On the inductive step, we need to construct $b_\alpha$. We first expand the language by adding new constant symbols $C = \{c_i : 0 \leq i < \alpha\}$ and interpret these symbols in $\mathcal{A}$ and $\mathcal{B}$ as $\{a_i : 0 \leq i < \alpha\}$ and $\{b_i : 0 \leq i \leq \alpha\}$ correspondingly. By the induction hypothesis, the expansions $\mathcal{A}_C$ and $\mathcal{B}_C$ are elementarily equivalent. Consider the type $\text{tp}_{\mathcal{B}_C}(b_\alpha)$ in the expanded language [that is, $\text{tp}(b_\alpha/C)$]. By saturatedness, this type is realised in $\mathcal{A}$, say by an element $a$, and we take $a$ to be $a_\alpha$. This satisfies the required property.

Finally, set $\pi : B \to A$ as $\pi : b_i \mapsto a_i$, and we are done by construction.

(ii) We use the previous method in combination with the back-and-forth procedure. Let

$$A = \{a_i : 0 \leq i \leq \kappa\}, \qquad B = \{b_i : 0 \leq i \leq \kappa\}$$

be the domains of saturated models $\mathcal{A}$ and $\mathcal{B}$ of cardinality $\kappa$, with ordinal orderings. We construct by induction on $\alpha < \kappa$ the subsets $A_\alpha \subset A$ and $B_\alpha \subset B$ with orderings

$$A_\alpha = \{a^j : j < \alpha\}, \qquad B_\alpha = \{b^j : j < \alpha\},$$

satisfying the conditions

$$\text{tp}(a^{j_1}, \ldots, a^{j_m}) = \text{tp}(b^{j_1}, \ldots, b^{j_m}) \tag{A.3}$$

for any finite sequences $0 \leq j_1 < \cdots < j_m < \alpha$;

$$\text{if } \delta + 2n < \alpha, \ \delta \text{ limit}, n \in \omega, \qquad \text{then } a_{\delta+n} \in A_\alpha \qquad (A.4)$$

$$\text{if } \delta + 2n + 1 < \alpha, \ \delta \text{ limit}, n \in \omega, \qquad \text{then } b_{\delta+n} \in B_\alpha. \qquad (A.5)$$

Clearly, Equation (A.3) implies that $a^j \mapsto b^j$ is an elementary monomorphism $A_\alpha \rightarrow B_\alpha$. When we reach $\alpha = \kappa$, this together with Equations (A.4) and (A.5) give us an isomorphism, $\mathcal{A} \cong \mathcal{B}$.

For $\alpha = 1$, take $a^0 := a_0$ and choose $b^0$ to be the first element among the $b_i$ satisfying the type $\mathrm{tp}(a^0)$.

Now assume that $A_\alpha$ and $B_\alpha$ have been constructed. We introduce constant symbols $c^j$, naming the $a^j$ in $\mathcal{A}$ and $b^j$ in $\mathcal{B}$. $C_\alpha = \{c^j : j < \alpha\}$.

If $\alpha$ is of the form $\delta + 2n$ and $a_{\delta+n} \notin A_\alpha$, we choose $a^\alpha := a_{\delta+n}$. If already $a_{\delta+n} \in A_\alpha$, we skip the step. Then we choose $b^\alpha$ to be the first element among $b_i$ satisfying the type $\mathrm{tp}(a^\alpha/C_\alpha)$. Such a $b_i$ does exist because card $C_\alpha < \kappa$ and $\mathcal{B}$ is $\kappa$-saturated.

If $\alpha$ is of the form $\delta + 2n + 1$ and $b_{\delta+n} \notin B_\alpha$, we choose $b^\alpha := b_{\delta+n}$. Then we choose $a^\alpha$ to be the first element among the $a_i$ satisfying the type $\mathrm{tp}(b^\alpha/C_\alpha)$.

In each case, Equations (A.3)–(A.5) are satisfied for $\alpha + 1$.

On limit steps $\lambda$ of the construction, we take

$$A_\lambda = \bigcup_{\alpha < \lambda} A_\alpha, \qquad B_\lambda = \bigcup_{\alpha < \lambda} B_\alpha.$$

This has the desired properties.

(iii) We use here another standard process.

*Claim.* Given a model $\mathcal{A}$ of $T$, there is an elementary extension $\mathcal{A}' \succeq \mathcal{A}$ such that any 1-type in $\mathrm{Th}(\mathcal{A})$ over any $C \subseteq A$, with card $C < \kappa$, is realised in $\mathcal{A}'$.

Indeed, by Exercise A.4.13, we can realise any set of types in some elementary extension of $\mathcal{A}$. This proves the claim.

Denote $\mathcal{A}$ as $\mathcal{A}^{(0)}$ and then construct, using the claim, an elementary chain of models

$$\mathcal{A}^{(0)} \preccurlyeq \mathcal{A}^{(1)} \preccurlyeq \cdots \preccurlyeq \mathcal{A}^{(\alpha)} \cdots$$

of length $\mu$ for a regular $\mu \geq \kappa$ (in particular $\mu = \kappa^+$ suffices in any case) such that $\mathcal{A}^{(\alpha+1)}$ realises all 1-types over subsets of $\mathcal{A}^{(\alpha)}$ of cardinality less than $\kappa$. Then the union $\mathcal{A}^* = \bigcup_{\alpha < \mu} \mathcal{A}^{(\alpha)}$ of the elementary chain, by Exercise A.2.9, is an elementary extension of $\mathcal{A}$ and indeed of each $\mathcal{A}^{(\alpha)}$. By the choice of $\mu$, for any subset $C$ of the domain $A^*$ of cardinality $< \kappa$, one can find $\lambda < \mu$ such that $C \subseteq \bigcup_{\alpha < \lambda} A^{(\alpha)} \subseteq A^{(\lambda)}$. It follows that $\mathcal{A}^*$ is a $\kappa$-saturated model of $T$. This proves (iii). $\qquad \square$

**Remark A.4.21.** The back-and-forth method used in the proof of (ii) is a universal tool in model theory, apparently first used by G. Cantor in his construction of the isomorphism between countable dense orders. In fact, Cantor's theorem is a special case of Theorem A.4.20(ii) because a dense linear order is $\omega$-saturated.

It follows from Theorem A.4.20(ii) that if $T_1$ and $T_2$ are complete theories in the same language, both having saturated models of the same cardinality, then $T_1 = T_2$ iff $\mathcal{A}_1 \cong \mathcal{A}_2$. This is a powerful criterion of elementary equivalence in a case when the existence of saturated models can be established. In general, a saturated model may not exist without assuming some form of generalised continuum hypothesis, but there are ways, using set-theoretic analysis, around this problem.

In fact, there is a way, less algebraic but quite universal, to apply a back-and-forth procedure to establish elementary equivalence.

**Definition A.4.22.** A back-and-forth system between $L$-structures $\mathcal{A}$ and $\mathcal{B}$ is a non-empty set $I$ of isomorphisms of sub-structures of $\mathcal{A}$ and sub-structures of $\mathcal{B}$ such that

$a \in \mathrm{Dom}\, f_0$ and $a' \in \mathrm{Range}\, f_0$, for some $f_0 \in I$, and
(forth) for every $f \in I$ and $a \in A$, there is a $g \in I$ such that $f \subseteq g$ and $a \in \mathrm{Dom}\, g$;
(back) for every $f \in I$ and $b \in B$, there is a $g \in I$ such that $f \subseteq g$ and $b \in \mathrm{Range}\, g$.

It is easy to prove the following theorem.

**Theorem (Ehrenfeucht–Fraisse criterion for saturated models).** *Given $\aleph_0$-saturated structures $\mathcal{A}$ and $\mathcal{B}$, $\mathcal{A} \equiv \mathcal{B}$ iff there is a back-and-forth system between the two structures.*

In fact, the existence of a back-and-forth system between $\mathcal{A}$ and $\mathcal{B}$ implies more than an elementary equivalence, that is, an equivalence for a first-order language. Recall that an $L_{\infty,\omega}$-language is the language which allows taking conjunctions and disjunctions of any family of $L_{\infty,\omega}$-formulas with a given finite number of free variables, as well as applying negation and usual quantifiers.

**Fact (C. Karp, in Kueker, 1975).** Two $L$-structures $\mathcal{A}$ and $\mathcal{B}$ are $L_{\infty,\omega}$-equivalent iff there is a back-and-forth system between the two structures.

**Remark A.4.23.** Note that this criterion can also be used to establish that the type of a tuple $a$ in a structure $\mathcal{A}$ is equal to that of $b$ in M. Just consider the extension with constants $c$ naming $a$ in one case and $b$ in the other. Then consider $\mathcal{A}$ as M with $c$ naming $a$ and $\mathcal{B}$ as M with $c$ naming $b$.

Now we return to discuss saturatedness.

**Exercise A.4.24.** *Prove that an algebraically closed field of infinite transcendence degree is saturated.*

Saturated structures play an important role in model theory. The reader familiar with algebraic geometry could compare it with the role played by a *universal domain* in the sense of A. Weil, that is, a field of infinite transcendence degree. In fact, it is convenient in a concrete context to fix a $\kappa$-saturated model *M of a given complete theory $T$, with a $\kappa$ 'large enough' (to all intents and purposes). Such a model is often called the **universal domain** for $T$. In model-theoretic slang, one more often refers to *M as the **monster model**.

**Definition A.4.25.** Given $C \subseteq A$, for $\mathcal{A} \models T$, we use $\mathcal{S}_n(C, T)$ to denote the set of all complete $n$-types of the theory $\mathrm{Th}(\mathcal{A}_C)$.

A theory $T$ is said to be $\kappa$-**stable**, for $\kappa \geq \aleph_0$, if $\mathrm{card}\,(\mathcal{S}_n(C, T)) \leq \kappa$, for any $C$ of cardinality less or equal to $\kappa$.

$T$ is said to be **stable** if it is $\kappa$-stable for some infinite $\kappa$.

**Theorem A.4.26.** *Assume $T$ is $\kappa$-stable. Then $T$ has a saturated model of cardinality $\kappa$.*

*Proof.* For $\kappa$ regular, one can prove the theorem by the construction in part (iii) of Theorem A.4.20. We can choose all $\mathcal{A}^{(\alpha)}$ to be of cardinality $\kappa$ and also $\mu = \kappa$. For singular cardinals, that is, when $\mathrm{cf}(\kappa) < \kappa$, the proof is more subtle. We skip it here. $\qquad\square$

$\aleph_0$-stability is traditionally referred to as $\omega$-stability. This is perhaps the most interesting case of $\kappa$-stability because this is the property of all countable theories categorical in uncountable cardinalities (see the next sub-section).

Suppose now that $T$ is a $\omega$-stable theory and $A$ is a countable subset of the monster model. Recall the following topological notions.

**Definition A.4.27.** For a topological space $X$, the **Cantor–Bendixson derivative** $d(X)$ is the subset of all the limit points in $X$.

Define by induction
$$d^0(X) = X$$
$$d^{\alpha+1}(X) = d(d^\alpha(X))$$
$$d^\lambda(X) = \bigcap_{\alpha < \lambda} d^\alpha(X) \qquad \text{for } \lambda \text{ limit.}$$

For compact $X$, $d^\alpha(X)$ is also compact. The ordinal $\alpha$, where the process is stabilised, $d^\alpha(X) = d^{\alpha+1}(X)$, is called the **Cantor–Bendixson rank** of $X$. Then $d^\alpha(X)$ is empty or perfect (the perfect kernel), that is, without isolated points.

In our situation by cardinality arguments in the Stone topology, the topological space $S_n(A)$ must have Cantor–Bendixson rank less than $\omega_1$ and the perfect kernel empty.

**Definition A.4.28.** For an $A$-definable formula (set) $\psi$, define $\mathrm{CB}(\psi)$ to be the Cantor–Bendixson rank of the Stone space:

$$S_n^\psi(A) = \{p \in S_n(A) : \psi \in P\}.$$

Define for a complete type $p \in S_n(A)$

$$\mathrm{CB}(p) = \max\{\mathrm{CB}(\psi) : \psi \in p\}.$$

By definition, for any $\psi$ there is only a finite number of types of CB-rank equal to that of $\psi$ and so there is a maximum for numbers $m$ such that $U$ has $m$ disjoint $A$-definable subsets of the same CB-rank. We call $\psi$ (Morley)-irreducible if $m = 1$. It follows from the definitions that for any $\psi$ over $A$, the definition of $\mathrm{CB}(\psi)$ can be given by induction as follows:

$\mathrm{CB}(\psi) \geq 1$ iff $\psi({}^*M) \neq \emptyset$

$\mathrm{CB}(\psi) \geq \alpha$ iff for any $\beta < \alpha$ there are infinitely many disjoint $A$-definable subsets of $\psi({}^*M)$ of CB-rank greater or equal to $\beta$.

**Definition A.4.29.** In an $\omega$-stable theory, Morley rank of a definable subset $U \subseteq {}^*M^n$ is defined as

$$\mathrm{rk}\,(U) = \mathrm{CB}(U) - 1$$

or equivalently by induction,

$\mathrm{rk}\,(U) \geq 0$ iff $U \neq \emptyset$

$\mathrm{rk}\,(U) \geq \alpha$ iff for any $\beta < \alpha$ there are infinitely many disjoint $A$-definable subsets of $U$ of Morley rank greater or equal to $\beta$.

## A.4.3 Categoricity in uncountable powers

The basis for the theory of categoricity in uncountable powers is the following theorem.

**Theorem A.4.30 (Ehrenfeucht–Mostowski).** *If a countable theory $T$ has infinite models, then for any infinite cardinal $\kappa$ there is a model M of $T$ such that* *for any $A \subseteq M$ the number of complete n-types over $A$ realised in* M *is of cardinality* card $A + \aleph_0$.

*Proof.* See the work of Marker (2002) or Hodges (1993). □

**Theorem A.4.31.** *If a countable theory $T$ is categorical in some uncountable cardinality $\kappa$, then $T$ is $\omega$-stable.*

*Proof.* Consider a countable subset $A$ of the universal domain *M. Assume a contradiction that $S_n(A)$ is uncountable for some $n$. In *M, all the types of $S_n(A)$ are realised, so there is a subset $D \subseteq {}^*M$ of cardinality $\aleph_1$ such that the set $S_n^D(A)$ of complete $n$-type over $A$ realised in $D$ is of cardinality $\aleph_1$. We may assume also $A \subset D$.

Let M be the unique model of $T$ of cardinality $\kappa$. By the Lowenheim–Skolem theorem, there is a model of $T$ of cardinality $\kappa$ with $D$ elementarily embedded in it. By categoricity, we can take this model to be M. We simply say $A \subseteq D \subseteq M$.

On the other hand, by Ehrenfeucht–Mostowski, we know that the unique model of cardinality $\kappa$ realises at most card $A + \aleph_0$ complete types over $A$. This contradicts the fact that $S_n^D(A)$ is uncountable. □

**Remark A.4.32.** Given an $\aleph_0$-saturated infinite model M of a $\omega$-stable theory $T$, there exists a definable set $U$ in M of Morley rank greater than 0. If rk $U > 1$, then by definition there must be a definable subset $U'$ of $U$ with $0 < \mathrm{rk}\, U' < \mathrm{rk}\, U$. By this argument, we can always find a definable $U$ of rank 1 that is irreducible. The structure on $U$ induced from M is by definition a **strongly minimal structure**. Strongly minimal structures are ubiquitous in models of $\omega$-stable theories. In the more special case of uncountably categorical theories, any strongly minimal sub-structure $U$ of a model M controls M in a very strong way, in particular the micro-geometry of $U$ affects the macro-geometry of M (see Section B.1.3).

The basic result of categoricity theory is the following. See the work of Marker (2002) and Hodges (1993) for proof.

**Theorem A.4.33 (M. Morley).** *If a countable theory $T$ is $\kappa$-categorical for some uncountable $\kappa$, then $T$ is categorical in all uncountable cardinals.*

We also state the following fundamental fact.

**Theorem A.4.34.** *In an uncountably categorical theory $T$ of countable language, Morley rank of any definable set is finite and satisfies the following:*

- rk $S = 0$ *iff $S$ is finite,*

- $\mathrm{rk}\,(S_1 \cup S_2) = \max\{\mathrm{rk}\,S_1, \mathrm{rk}\,S_2\}$,
- *for the projection* $\mathrm{pr} : M^n \to M^k$

$$\mathrm{rk}\,S \le \mathrm{rk}\,\mathrm{pr}\,(S) + \max_{t \in \mathrm{pr}(S)} \mathrm{rk}\,\mathrm{pr}^{-1}(t) \cap S, \text{ and}$$

- *suppose* $\mathrm{rk}\,\mathrm{pr}^{-1}(t) \cap S$ *is the same for all* $t \in \mathrm{pr}(S)$; *then*

$$\mathrm{rk}\,S = \mathrm{rk}\,\mathrm{pr}\,(S) + \mathrm{rk}\,\mathrm{pr}^{-1}(t) \cap S.$$

*Proof.* See the work of Baldwin (1973) or Zilber (1974). We also give a proof of this theorem in the special case when $T$ is the theory of a strongly minimal structure in Lemma B.1.26.                                      $\square$

# Appendix B

## Elements of geometric stability theory

### B.1 Algebraic closure in abstract structures

**Definition B.1.1.**

$$\mathrm{acl}(A) = \{b \in M : \text{ there are } a \in A^n, m \in \mathbb{N}, \text{ and } \varphi(u, v) \text{ such that}$$

$$M \models \varphi(a, b) \, \& \, \exists^{\leq m} v \varphi(a, v)\}.$$

**Exercise B.1.2.** *The following properties of* acl *hold in any structure:*

$$A \subseteq B \text{ implies } A \subseteq \mathrm{acl}(A) \subseteq \mathrm{acl}(B) \tag{B.1}$$

$$\mathrm{acl}(\mathrm{acl}(A)) = \mathrm{acl}(A). \tag{B.2}$$

For any field $K$ and $A \subseteq K$, $\mathrm{acl}(A)$ contains the field-theoretic algebraic closure of $A$ in $K$.

**Lemma B.1.3.** *Any elementary monomorphism $\alpha$ between $A, A' \subseteq M$ can be extended to* $\mathrm{acl}(A) \to \mathrm{acl}(A')$.

*Proof.* Enumerate $\mathrm{acl}(A)$ and go by transfinite induction extending $\alpha$ to $A \cup \{a_i : i < \gamma\}$ finding for $a_\gamma$ the corresponding element $a'_\gamma \in \mathrm{acl}(A')$ as a realisation of type

$$\alpha(\mathrm{tp}(a_\gamma/(A \cup \{a_i : i < \gamma\}))$$

obtained by replacing elements of $\mathrm{acl}(A)$ by corresponding elements of $\mathrm{acl}(A')$ in $\mathrm{tp}(a_\gamma/(A \cup \{a_i : i < \gamma\}))$. Both types are principal and algebraic because $a_\gamma$ is algebraic over $A \cup \{a_i : i < \gamma\}$. Notice that after exhausting the process on $\mathrm{acl}(A)$, the other part, $\mathrm{acl}(A')$, is exhausted too, because going back from $\mathrm{acl}(A')$ to $\mathrm{acl}(A)$ we find elements in $\mathrm{acl}(A)$. $\square$

**Minimal structures**

**Definition B.1.4.** A structure M is said to be **minimal** if any subset of $M$ definable using parameters is either finite or a complement of a finite one.

We assume everywhere that the language of M is countable.

**Lemma B.1.5.** *In minimal structures, the following* **exchange principle** *holds:*

*For any* $A \subseteq M$, $b, c \in M$ : $b \in \mathrm{acl}(A, c) \setminus \mathrm{acl}(A) \rightarrow c \in \mathrm{acl}(A, b)$.

$$(B.3)$$

*Proof.* Suppose $b \in \mathrm{acl}(A, c) \setminus \mathrm{acl}(A)$. Then for some $\varphi(x, y)$ over $A$ and some $m$,

$$M \models \varphi(b, c) \ \& \ \exists^{\leq m} x \, \varphi(x, c). \tag{B.4}$$

W.l.o.g., we assume

$$M \models \varphi(x, y) \rightarrow \exists^{\leq m} x \, \varphi(x, y). \tag{B.5}$$

Suppose, towards a contradiction, that $\varphi(b, M)$ is infinite. Then $\mathrm{card}\,(\neg\varphi(b, M)) \leq k$ for some $k$ [i.e. M $\models \exists^{\leq k} y \neg \varphi(b, y)$], and

$$B = \{b' \in M : M \models \exists^{\leq k} y \neg \varphi(b', y)\}$$

is infinite, because $b \notin \mathrm{acl}(A)$. Choose distinct $b_1, \ldots, b_{m+1} \in B$. Then

$$\varphi(b_1, M) \cap \cdots \cap \varphi(b_{m+1}, M)$$

is infinite and thus contains a point $c'$. It contradicts Equation (B.5). $\square$

## B.1.1 Pre-geometry and geometry of a minimal structure

**Definition B.1.6.** An [abstract] **pre-geometry** is a set $M$ with an operator

$$\mathrm{cl} : 2^M \rightarrow 2^M$$

of finite character, that is, for any $A \subseteq M$ : $\mathrm{cl}(A) = \{\mathrm{cl}(A') : A' \subseteq A \text{ finite}\}$ and satisfying the conditions (1)–(3).

A pre-geometry is said to be a **geometry** if

$$\text{for any } a \subseteq M, \qquad \mathrm{cl}(\{a\}) = \{a\}. \tag{B.6}$$

**Getting a geometry from a pre-geometry**

**Lemma B.1.7.** *The relation* $\sim$ *on* $M \setminus \mathrm{cl}(\emptyset)$ *defined as*

$$x \sim y \qquad \text{iff } \mathrm{cl}(x) = \mathrm{cl}(y)$$

*is an equivalence relation.*

*Proof.* The proof follows from the exchange principle. □

Define for a pre-geometry M the set

$$\hat{M} = (M \setminus \text{cl}(\emptyset))/ \sim .$$

Then any point in $\hat{M}$ is of the form $\hat{a} = \text{cl}(a) \setminus \text{cl}(\emptyset)$ for a corresponding $a \in M \setminus \text{cl}(\emptyset)$. For a subset $\hat{A} = \{\hat{a} : a \in A\} \subseteq \hat{M}$, define

$$\text{cl}(\hat{A}) = \{\hat{b} : b \in \text{cl}(A)\}.$$

The operator cl on $\hat{M}$ then satisfies (1)–(4), and thus $\hat{M}$ is a geometry.

A pre-geometry M with a fixed $D \subseteq M$ gives rise to another pre-geometry, $M_D$, **the localisation of** M **with respect to** $D$: the set of $M_D$ is just $M$ and $\text{cl}_D(A) = \text{cl}(D \cup A)$.

**Subspaces** of a pre-geometry are subsets of the form $\text{cl}(A)$. Pre-geometry is said to be **locally finite** if $\text{cl}(A)$ is finite whenever $A$ is.

**Example B.1.8.** *Vector spaces over division rings are pre-geometries if we let*

$$\text{cl}(A) = \text{span}(A).$$

**Projective space** *associated with a vector space* M *is defined exactly as the geometry* $\hat{M}$. **Affine space** *associated with a vector space* M *is defined on the same set* M *by the new closure operator:*

$$\text{cl}_{\text{aff}}(A) = A + \text{span}(A - A)$$

*where* $A - A = \{a_1 - a_2 : a_1, a_2 \in A\}$.

**Exercise B.1.9.** *Show that an affine space is a geometry and its localisation with respect to any point is isomorphic to the initial vector space pre-geometry.*

**Definition B.1.10.** A set $A$ is said to be **independent** if $\text{cl}(A) \neq \text{cl}(A')$ for any proper subset $A' \subset A$.

A maximal independent subset of a set $A$ is said to be a **basis of** $A$.

**Lemma B.1.11.** *Any two bases $B$ and $C$ of a set $A$ are of the same cardinality.*

*Proof.* First consider the case when $B$ is finite and consists of $n$ elements $b_1, \ldots, b_n$. There is a $c \in C$ such that

$$c \in \text{cl}(b_1, \ldots, b_n) \setminus \text{cl}(b_1, \ldots, b_{n-1}),$$

because otherwise $B$ is not independent. By the exchange principle, $\{c, b_1, \ldots, b_{n-1}\}$ is a basis of $A$. In the localisation $M_c$, the sets $\{b_1, \ldots, b_{n-1}\}$ and $C \setminus \{c\}$ are bases of $A$. By induction on $n$, the statement follows.

Consider now the case when both $B$ and $C$ are infinite. It follows from the finite character of cl that for any $b \in B$ there is a minimal finite $C_b \subset C$ such that $b \in \mathrm{cl}(C_b)$. Thus there is a mapping of $B$ into $\mathcal{P}_{\mathrm{fin}}(C)$, the set of all finite subsets of $C$. The mapping is finite-to-one, because by the previous argument proved that the set

$$\{d \in B : C_d = C_b\}$$

is an independent subset of $\mathrm{cl}(C_b)$ and proved its size is not bigger than the size of $C_b$.

It follows that card $B \leq$ card $C$. By the symmetry, card $B =$ card $C$. $\qquad \square$

Now we can give the following definition.

**Definition B.1.12.** For any subset $A$ of a pre-geometry, we define the (**combinatorial**) **dimension** cdim $A$ to be the cardinality of a basis of $A$. If also $B \subseteq A$, then $\mathrm{cdim}\,(A/B)$ is the dimension of $A$ in the pre-geometry $\mathbf{M}_B$.

**Remark B.1.13.** There are many notions of a dimension in mathematics, including the notion dim $V$ of the dimension of an algebraic variety $V$, and stability theory treats and compares them in a systematic way.

**Lemma B.1.14 (addition formula).**

$$\mathrm{cdim}\,(A/B) + \mathrm{cdim}\,(B) = \mathrm{cdim}\,(A).$$

*Proof.* One can construct a basis of $A$ by adjoining to a basis of $B$ a basis of $A$ in $\mathbf{M}_B$. $\qquad \square$

**Example B.1.15.** *The transcendence degree of a subset $A$ of an algebraically closed field $F$ is just* cdim $A$, *which is well defined because $F$ is a minimal structure (with* cl $=$ acl).) *Because any field is a sub-field of an algebraically closed one, the definition is applicable for subsets of any field.*

**Lemma B.1.16.** *For $X, Y \subseteq M$ subspaces of a pre-geometry,*

$$\mathrm{cdim}\,(X \cup Y) \leq \mathrm{cdim}\,X + \mathrm{cdim}\,Y - \mathrm{cdim}\,(X \cap Y).$$

*Proof.* Let $Z$ be a basis of $X \cap Y$. Let $Z \cup X_0$ and $Z \cup Y_0$ be bases of $X$ and $Y$, respectively. Then $\mathrm{cl}(X_0 \cup Z \cup Y_0) = \mathrm{cl}(X \cup Y)$, and thus $\mathrm{cdim}\,(X \cup Y) \leq |X_0 \cup Z| + |Z \cup Y_0| - |Z|$. $\qquad \square$

**Definition B.1.17.** A subset $A$ of a structure M is said to be **indiscernible over** $B$ if $\mathrm{tp}(\bar{a}/B) = \mathrm{tp}(\bar{a}'/B)$ for any two $n$-tuples of distinct elements of $A$ for any finite $n$.

**Proposition B.1.18.** *Let* M *be a minimal structure,* $A, B \subseteq M$, *and* A *be independent over* B *(in the pre-geometry of* M *). Then* A *is indiscernible over* B. *Moreover, the* n-*type* $\mathrm{tp}(\bar{a}/B)$ *for* $\bar{a} \in A^n$ *does not depend on* B.

*Proof.* Consider $\bar{a} = \langle a_1, \ldots, a_n \rangle$, $\bar{a}' = \langle a_1', \ldots, a_n' \rangle$ all with distinct co-ordinates from $A$. The size $n = 1$ $\mathrm{tp}(a/B)$ is just the set of those formulas $\varphi(x)$ over $B$ which have $\varphi(M)$ infinite. The same characterises $\mathrm{tp}(a'/B)$, and thus the types are equal.

For $n > 1$, suppose as an inductive hypothesis that the tuples have the same type over $B$. Then for $a_{n+1} \in A \setminus \{a_1, \ldots a_n\}$ and any formula $\varphi(\bar{x}, y)$ over $B$,

$$\models \varphi(\bar{a}, a_{n+1}) \qquad \text{iff} \qquad \varphi(\bar{a}, M) \text{ is infinite.}$$

By the equality of types, the latter is equivalent to $\varphi(\bar{a}', M)$ and is infinite which gives $\models \varphi(\bar{a}', a_{n+1}')$ for any $a_{n+1}' \in A$ distinct from the co-ordinates of $\bar{a}'$. $\square$

**Lemma B.1.19.** *Any minimal structure in a countable language is homogeneous.*

*Proof.* Suppose $B, B' \subseteq M$ are of cardinality less than $\mathrm{card}\, M$ and there is an elementary monomorphism $\alpha : B \to B'$. It follows that $\mathrm{cdim}\, B = \mathrm{cdim}\, B'$. From the assumptions on cardinalities, it follows also that $\mathrm{cdim}\, B < \mathrm{card}\, M$. Thus, from the addition formula, $\mathrm{cdim}\, M/B = \mathrm{cdim}\, M/B'$. Let $A \supseteq B$ and $A' \supseteq B'$ be extensions to bases of $M$ over $B$ and $B'$, respectively. Because

$$\mathrm{card}\,(A \setminus B) = \mathrm{card}\,(A' \setminus B'),$$

there is a bijection $\beta : A \to A'$ extending $\alpha$. The description of types of $n$-tuples in bases shows that $\beta$ is an elementary monomorphism. Now Lemma B.1.3 finishes the proof. $\square$

## B.1.2 Dimension notion in strongly minimal structures

**Theorem B.1.20.** *Minimal structures of infinite dimension are saturated.*

*Proof.* Notice that in this case $\mathrm{cdim}\, M = \mathrm{card}\, M$. Let $\varphi(x)$ be a formula over $A$, $\mathrm{card}\, A < \mathrm{card}\, M$. Either $\varphi(M)$ is finite or $M \setminus \mathrm{cl}(A) \subseteq \varphi(M)$. Thus, for any consistent set of such formulas,

$$\bigcap_i \varphi_i(M)$$

either contains the non-empty set $M \setminus \mathrm{cl}(A)$ or is a non-empty subset of some finite $\varphi_i(M)$. $\square$

**Corollary B.1.21.** *Any structure which is elementarily equivalent to a minimal one of infinite dimension is minimal too. It also satisfies the* **finite cover property** **(f.c.p.)**: *for any* $\varphi(x, \bar{y})$, *there is a natural number m such that* card $\varphi(M, \bar{a}) > m$ *implies* $\varphi(M, \bar{a})$ *is infinite.*

**Definition B.1.22.** A minimal structure is said to be **strongly minimal** if it is elementarily equivalent to a minimal structure of infinite dimension.

### Rank notion for sets definable in strongly minimal structures

We assume $M$ is strongly minimal of infinite dimension.

**Definition B.1.23.** Let $A \subseteq M$ be finite and M be saturated minimal. For an $A$-definable subset $S \subseteq M^n$, put the **Morley rank** as

$$\text{rk } S = \max_{\langle s_1, \ldots, s_n \rangle \in S} \text{cdim}(\{s_1, \ldots, s_n\}/A).$$

**Lemma B.1.24.** rk $\varphi(M)$ *has the same value in every saturated structure elementarily equivalent to a given strongly minimal one and does not depend on A.*

*Proof.* The proof is immediate from the saturation of M.                    □

**Definition B.1.25.** For an arbitrary strongly minimal structure M, rk $\varphi(M)$ is defined as the rank in saturated elementary extensions of M.

**Lemma B.1.26 (basic rank properties).** *For any strongly minimal structure M, the following statements are true.*

*(i)* rk $M^n = n$.
*(ii)* rk $S = 0$ *iff S is finite.*
*(iii)* rk $(S_1 \cup S_2) = \max\{\text{rk } S_1, \text{rk } S_2\}$.
*(iv)* *For the projection* pr $: M^n \to M^k$,

$$\text{rk } S \leq \text{rk pr}(S) + \max_{t \in \text{pr}(S)} \text{rk pr}^{-1}(t) \cap S.$$

*(v)* *Suppose* rk pr$^{-1}(t) \cap S$ *is the same for all* $t \in \text{pr}(S)$. *Then*

$$\text{rk } S = \text{rk pr}(S) + \text{rk pr}^{-1}(t) \cap S.$$

*Proof.* Statements (i)–(iii) are immediate from the definition.
(iv) Let $\langle s_1, \ldots s_n \rangle \in S$ be of maximal dimension in $S$. Then

$$\text{rk } S = \text{cdim}(\{s_1, \ldots, s_n\}/A) = \text{cdim}(\{s_1, \ldots, s_n\}/\{s_1, \ldots, s_k\} \cup A)$$
$$+ \text{cdim}(\{s_1, \ldots, s_k\}/A) \leq \text{rk pr}^{-1}(\langle s_1, \ldots, s_k \rangle) \cap S + \text{rk pr } S.$$

(v) If one chooses first a tuple $\langle s_1, \ldots, s_k \rangle \in \mathrm{pr}\, S$ of maximal possible dimension and then extends it to $\langle s_1, \ldots, s_n \rangle \in S$ of maximal possible dimension over $\{s_1, \ldots, s_k\} \cup A$, then

$$\mathrm{cdim}\,(\{s_1, \ldots, s_n\}/\{s_1, \ldots, s_k\} \cup A) = \mathrm{rk}\, \mathrm{pr}^{-1}(\langle s_1, \ldots, s_k \rangle),$$
$$\mathrm{cdim}\,(\{s_1, \ldots, s_k\}/A) = \mathrm{rk}\, \mathrm{pr}\, S,$$

and thus

$$\mathrm{rk}\, S \geq \mathrm{rk}\, \mathrm{pr}\, S + \mathrm{rk}\, \mathrm{pr}^{-1}(\langle s_1, \ldots, s_k \rangle).$$

□

**Lemma B.1.27.** *For any definable $S \subseteq M^n$, there is an upper bound on $m \in \mathbb{N}$ such that $S$ can be partitioned into $k$ disjoint subsets*

$$S = S_1 \cup \cdots \cup S_m,$$

*each of rank equal to* $\mathrm{rk}\, S$.

*Proof.* We use induction on $n$. For $n = 1$, the statement follows from the definitions.

For arbitrary $n$, let $\mathrm{rk}\, S = k$. This means there is a point $\langle s_1, \ldots, s_n \rangle \in S$ of dimension $k$, and thus some $\{s_{i_1}, \ldots, s_{i_k}\}$ are independent. Let us consider the case $\langle i_1, \ldots, i_k \rangle = \langle 1, \ldots, k \rangle$.

Then $s_j \in \mathrm{cl}\{s_1, \ldots, s_k\}$ for all $j = 1, \ldots, n\}$; thus for some natural number $l = l_{i_1, \ldots, i_k}$,

$$\models \exists^{=l} \langle x_1, \ldots, x_n \rangle \in S : \langle x_1, \ldots, x_k \rangle = \langle s_1, \ldots, s_k \rangle.$$

Note the formula $\psi(\langle s_1, \ldots, s_k \rangle)$, and notice that $\psi(M) \subseteq M^k$ is of rank $k$. Let

$$S^0 = \{\langle s_1, \ldots, s_n \rangle \in S : \psi(\langle s_1, \ldots, s_k \rangle)\}.$$

By these formulas, $\mathrm{rk}\, S^0 = \mathrm{rk}\, \psi(M) = \mathrm{rk}\, S$. Suppose

$$S^0 = S_1^0 \cup \cdots \cup S_m^0$$

is a partition and all the summands are $A'$-definable of rank $k$. Then necessarily for any $j \leq m$, there is $\langle s_{j,1}, \ldots, s_{j,n} \rangle \in S_j^0$ with the first $k$ co-ordinates independent over $A'$. By indiscernibility, we can choose $\langle s_{j,1}, \ldots, s_{j,n} \rangle \in S_j^0$ so that

$$\langle s_{j,1}, \ldots, s_{j,k} \rangle = \langle s_{1,1}, \ldots, s_{1,n} \rangle$$

for all $j$. It follows immediately that $m \leq l$.

Taking into account all possibilities for $\langle i_1, \ldots, i_k \rangle$, we get

$$m \leq \sum_{\{i_1, \ldots, i_k\}} l_{i_1, \ldots, i_k}.$$

$\square$

**Definition B.1.28.** The exact upper bound for an equirank partition of $S$ is called the **Morley degree of** $S$ and denoted $\mathrm{Mdeg}(S)$. A definable set of Morley degree 1 is called **(Morley) irreducible**.

**Definition B.1.29.** For a type $p(\bar{x})$ definable over $A$, the Morley rank of type is defined as

$$\mathrm{rk}\,(p) = \min\{\mathrm{rk}\,\varphi(\bar{x}) : \varphi \in p\}.$$

For a point $\bar{s} \in M^n$ and a subset $A \subseteq M$, the Morley rank of the point over $A$ is defined as

$$\mathrm{rk}\,(\bar{s}/A) = \mathrm{rk}\,(\mathrm{tp}(\bar{s}/A)).$$

A point $\bar{s} \in S \subseteq M^n$ for an irreducible subset $S$ defined over $A$ is said to be **generic over** $A$ if

$$\mathrm{rk}\,(\bar{s}/A) = \mathrm{rk}\,S.$$

**Lemma B.1.30.** *For an irreducible $S$ over $A$, there is a unique complete type $p$ over $A$ containing $S$. More exactly,*

$$p = \mathrm{tp}(s/A)$$

*for $s$ generic in $S$. In particular, any two generic points have the same type over $A$.*

*Proof.*

$$p = \{\varphi(\bar{x}) \text{ over } A : \mathrm{rk}\,(\varphi(M) \cap S) = \mathrm{rk}\,S\}.$$

The rest follows from definitions. $\square$

**Lemma B.1.31 (addition formula for tuples).**

$$\mathrm{rk}\,(\bar{b}\bar{c}/A) = \mathrm{rk}\,(\bar{b}/A\bar{c}) + \mathrm{rk}\,(\bar{c}/A).$$

*Here $A\bar{c} = A \cup |\bar{c}|$, where $|\bar{c}|$ is the set of the co-ordinates of $\bar{c}$.*

*Proof.* The proof follows from the addition formula for dimensions taking into account that $\operatorname{rk}(\bar{b}/A) = \operatorname{cdim}(|\bar{b}|/A)$, which follows immediately from the definitions. □

**Definition B.1.32.** Two points $\bar{b} \in M^k$ and $\bar{c} \in M^n$ are said to be **independent over** $A$ if

$$\operatorname{rk}(\bar{b}/A\bar{c}) = \operatorname{rk}(\bar{b}/A).$$

**Lemma B.1.33.** *The independence relation is symmetric.*

*Proof.* $\operatorname{rk}(\bar{b}\bar{c}/A) = \operatorname{rk}(\bar{b}/A\bar{c}) + \operatorname{rk}(\bar{c}/A) = \operatorname{rk}(\bar{c}/A\bar{b}) + \operatorname{rk}(\bar{b}/A)$ by the addition formula. Then if $\operatorname{rk}(\bar{b}/A\bar{c}) = \operatorname{rk}(\bar{b}/A)$, so $\operatorname{rk}(\bar{c}/A\bar{b}) + \operatorname{rk}(\bar{c}/A)$. □

**Lemma B.1.34 (definability of Morley rank).** *For any formula $\varphi(\bar{x}, \bar{y})$ with* length$(\bar{x}) = k$, length$(\bar{y}) = n$, *and any* $m$, *the set*

$$\{\bar{a} \in M^k : \operatorname{rk}\varphi(\bar{a}, M) \geq m\}$$

*is definable.*

*Proof.* The proof follows by induction on $n$. For $n = 1$, $\operatorname{rk}\varphi(a, M) \geq 0$ iff $\varphi(a, M) \neq \emptyset$, and $\operatorname{rk}\varphi(a, M) \geq 1$ iff $\varphi(a, M)$ is infinite iff $\operatorname{card}\varphi(a, M) \geq n_\varphi$ by f.c.p.

For arbitrary $n$,

$$\operatorname{rk}\varphi(a, y_1, \ldots, y_n) \geq m \qquad \text{iff} \qquad \{b \in M : \operatorname{rk}\varphi(a, b, y_2, \ldots, y_n) \geq m - 1\}$$

is infinite or

$$\{b \in M : \operatorname{rk}\varphi(a, b, y_2, \ldots, y_n) \geq m\} \neq \emptyset$$

by the addition formula for ranks. Both conditions on the right-hand side are definable by the induction hypothesis. □

## Sets definable in M

We consider Morley rank for sets definable in strongly minimal M. Recall that any such set is of the form $U = S/E$, where $S \subseteq M^n$ is a definable subset and $E \subseteq S^2 \subseteq M^{2n}$ is a definable subset which is an equivalence relation. We consider only $U$ such that $E$ is equirank [i.e. $\operatorname{rk}E(s, M)$ is of the same value for all $s \in S$].

**Definition B.1.35.**

$$\operatorname{rk}U = \operatorname{rk}S - \operatorname{rk}E(s, M) \text{ for } s \in S.$$

**Lemma B.1.36.** *The definition is invariant under definable bijections; that is, if there is a bijection*

$$f : S_1/E_1 \to S_2/E_2$$

*and $f$ is a definable function, then* rk $S_1/E_1 =$ rk $S_2/E_2$.

*Proof.* By definition $f = F/E$, where $F \subseteq S_1 \times S_2$, $E = E_1 \times E_2$, and the following hold, for any $s_1, s_1' \in S_1$, $s_2, s_2' \in S_2$,

$$F(s_1, s_2) \;\&\; F(s_1', s_2') \to (E_1(s_1, s_1') \leftrightarrow E_2(s_2, s_2')),$$

$$\text{pr}_{S_1} F = S_1 \qquad \text{and} \qquad \text{pr}_{S_2} F = S_2.$$

From the addition formula, by projecting on $S_1$, we get

$$\text{rk } F = \text{rk } S_1 + \text{rk } E_2(s_2, M),$$

and by projecting on $S_2$, we get

$$\text{rk } F = \text{rk } S_2 + \text{rk } E_1(s_1, M).$$

It follows that

$$\text{rk } S_1 - \text{rk } E_1(s_1, M) = \text{rk } S_2 - \text{rk } E_2(s_2, M).$$

$\square$

**Proposition B.1.37.** *Basic rank properties (i)–(v) hold for definable sets, as well as for Lemma B.1.27 and the definability of rank.*

*Proof.* The proof is done by appropriately using the same arguments as in the proofs of the statements mentioned. $\square$

**Proposition B.1.38 (finite equivalence relation theorem).** *For any $A$-definable set $S$ of rank $k$, there is an $A$-definable subset $S^0 \subseteq S$ and an equivalence relation $E$ on $S^0$ such that $S^0 \sqsubset\!\!\!\sqsupset S$, $S^0/E$ is finite, and each equivalence class is of rank $k$ and irreducible.*

We omit the proof of the theorem, which can be found elsewhere.

### B.1.3 Macro- and micro-geometries on a strongly minimal structure

The geometry defined on a strongly minimal $M$ in Definition B.1.1 would be called the **micro-geometry** on $M$ as opposed to the **macro-geometry** that is induced by M under certain conditions, as constructed in the following proposition.

**Proposition B.1.39.** *Suppose there are $a_1, a_2, b_1, b_2, c \in M$, every four of which are independent, $c \in cl(a_1, a_2, b_1, b_2)$ and*

$$cl(a_1, a_2, c) \cap cl(b_1, b_2, c) = cl(c).$$

*Then an incidence system $(S, P, I)$ is definable in M with properties*

$$rk(S) = 2, \qquad Mdeg(S) = 1 \ \ rk(L) \geq 2,$$

$$rk(Ip) = 1 \qquad \textit{for all } p \in P,$$

*if $p_1, p_2 \in P$, $\quad p_1 \neq p_2$, $\qquad$ then $Ip_1 \cap Ip_2$ is finite or empty.*

*Proof.* $S_0 = M \times M$, $P_0 = M \times M \times M$, and let $I_0 \subseteq S_0 \times P_0$ be an $\emptyset$-definable relation such that

$$\langle b_1, b_2 \rangle I_0 \langle a_1, a_2, c \rangle$$

$$\langle x_1, x_2 \rangle I_0 \langle y_1, y_2, z \rangle \rightarrow z \in cl(x_1, x_2, y_1, y_2).$$

A relation witnessing the dependence among $a_1, a_2, b_1, b_2, c$ has these properties.

Let $p_0 = \langle a_1, a_2, c \rangle$. Then $I_0 p_0$ is an $p_0$-definable set of Morley rank 1. By the finite equivalence relation theorem, using $p_0$ one can define an equivalence relation $E_{p_0}$ on $I_0 p_0$ with finitely many classes, and, say, $m$ of them of rank 1 irreducible.

Let

$$I_1 = \{\langle s, p \rangle \in I_0 : E_p \text{ is an equivalence relation with exactly } m \text{ infinite}$$
$$\text{classes and } s \text{ is in one of them}\}.$$

Define the binary relation $E$ on $I_1$:

$$\langle s, p \rangle E \langle s', p' \rangle \qquad \text{iff } p = p' \ \& \ s E_p s'.$$

Define

$$P_1 = I_1/E,$$

and for $q \in P_1$, $s \in S_0$, write $s I_2 q$ iff $q = \langle s, p \rangle$ for some $p \in P_0$.

By definitions there is a canonical mapping,

$$\alpha : P_1 \rightarrow P_0,$$

corresponding to the projection $I_1 \rightarrow P_0$, which is exactly $m$-to-one mapping. Also, for all $q \in P_1$,

$$\operatorname{rk}(I_2 q) = 1.$$

By definitions, for $q_0$ corresponding to $p_0$ via $\alpha$, $I_2 q_0$ is irreducible. Define

$$P_2 = \{q \in P_1 : \forall q' \in P_1 \operatorname{rk}(I_2 q \cap I_2 q') = 1 \rightarrow I_2 q \sqsubset I_2 q'\}.$$

It follows from the remark that $q_0 \in P_2$ and for all $q \in P_2$, $\operatorname{rk}(I_2 q) = 1$. Define an equivalence relation on $P_2$ of

$$q F q' \qquad \text{iff} \qquad I_2 q \sqsubset\!\!\sqsupset I_2 q'.$$

We are now in the situation of claim 2 of the proof of the preceding theorem. It follows that

$$q F \sqsubset s I_2$$

whenever $s$ is generic in $I_2 q$ over $q$ and $s$ is generic in $S_0$. Define

$$P_3 = P_2/F$$

and for $\bar{p} \in P_3$, $s \in S_0$

$$s I_3 \bar{p} \qquad \text{iff} \qquad \bar{p} \sqsubset I_2 s.$$

This proves $s_0 I_3 \bar{p}_0$ holds, where $\bar{p}_0$ is obtained throughout the construction from $p_0$, $s_0$. Also, by the construction $\bar{p}_0 \in \operatorname{cl}(p_0)$.

Because $s_0 \in I_3 \bar{p}_0$, it follows that $\operatorname{rk}(I_3 \bar{p}_0) \geq 1$. On the other hand, if $s \in I_3 \bar{p}_0$ is of maximal rank over $\bar{p}_0$ and $q \in p_0 F$ is of maximal rank over $s$, $\bar{p}_0$, then by definition $s I_2 p$ holds and $q$ and $s$ are independent over $\bar{p}_0$. It follows that $\operatorname{rk}(s/\bar{p}_0) = \operatorname{rk}(s/\bar{p}_0, q) \leq 1$. Thus

$$\operatorname{rk}(I_3 \bar{p}_0) = 1.$$

Let

$$P = \{\bar{p} \in P_3 : \operatorname{rk}(I_3 \bar{p}_0) = 1\}, \qquad S = \{s \in S_0 : \exists \bar{p} \in P \, s I_3 \bar{p}\},$$

$$I = I_3 \cap (S \times P).$$

Now we need to show that for $\bar{p}_1$, $\bar{p}_2$ distinct from $P$, $I \bar{p}_1 \cap I \bar{p}_2$ is finite. So, suppose $s$ is a point in the intersection. Choose $\langle q_1, q_2 \rangle \in p_1 F \times p_2 F$ of maximal rank over $s$, $\bar{p}_1$, $\bar{p}_2$. Then $s \in I_2 q_1 \cap I_2 q_2$ and $s$ is independent with $q_1, q_2$ over $\bar{p}_1, \bar{p}_2$. Then

$$\text{rk}\,(s/\bar{p}_1, \bar{p}_2) = \text{rk}\,(s/\bar{p}_1, \bar{p}_2, q_1, q_2) < 1$$

because $\neg q_1 F q_2$.

To finish the proof, we need to show that $\text{rk}\,(P) \geq 2$ which would follow from $\text{rk}\,(\bar{p}_0/\emptyset) \geq 2$.

Suppose towards the contradiction $\text{rk}\,(\bar{p}_0/\emptyset) \leq 1$. Then, because $\text{rk}\,(s_0/\bar{p}_0) = 1 < \text{rk}\,(s_0/\emptyset)$, we have $\text{rk}\,(\bar{p}_0/s_0) < \text{rk}\,(\bar{p}_0/\emptyset)$; that is, $\bar{p}_0 \in \text{cl}(s_0) = \text{cl}(b_1, b_2)$. Then, from the assumptions of the proposition, $c \notin \text{cl}(\bar{p}_0)$.

On the other hand, $\bar{p}_0 \in \text{cl}(p_0) = \text{cl}(a_1, a_2, c)$. It follows $b_1 \notin \text{cl}(p_0)$, $b_2 \notin \text{cl}(p_0)$. Therefore, there exists $c' \in M$ such that

$$\text{tp}(cc'/\bar{p}_0) = \text{tp}(b_1 b_2/\bar{p}_0) = \text{tp}(s_0/\bar{p}_0).$$

Thus, $\text{rk}\,(cc'/\bar{p}_0) = 1$ and so

$$c' \in \text{cl}(\bar{p}_0, c) \subseteq \text{cl}(a_1, a_2, c) \cap \text{cl}(b_1, b_2, c).$$

Hence $\text{cl}(c') = \text{cl}(c)$, which contradicts $\text{cl}(b_1) \neq \text{cl}(b_2)$. □

**Definition B.1.40.** It is said, with a slight deviation from the standard terminology, that a **pseudo-plane** is definable in M if there is a two-sorted structure $(S, P, I)$ definable in $M_A$, some $A$, with properties stated in the proposition.

**Definition B.1.41.** An (abstract) **projective geometry** is a set of 'points' and 'lines' satisfying these conditions:

(i) through any two points there is a line;
(ii) there are at least three points on every line;
(iii) two distinct lines intersect in at most one point; and
(iv) for any distinct points $a, b, c, d$, if lines $(a, b)$ and $(c, d)$ intersect then lines $(a, c)$ and $(b, d)$ do.

The geometry M is said to be **locally projective** if for generic $c \in M$, the geometry $\hat{M}_c$ is isomorphic to a projective geometry over a division ring.

Any three points $a, b, c$ of a projective geometry which do not lie on a common line **generate a projective plane** as the set of points

$$S(a, b, c) = \bigcup \{(a, z) : z \in (b, c)\}.$$

By condition (iv), the plane generated by any non-collinear $a', b', c' \in S(a, b, c)$ coincides with $S(a, b, c)$. The $n$-**subspaces** of a projective geometry are defined by induction as

$$S(a_1, \ldots, a_{n+1}) = \bigcup \{(a_{n+1}, z) : z \in (a_1, \ldots, a_n)\}$$

for $a_1, \ldots, a_{n+1}$ not in an $(n - 1)$-subspace. Again by axiom (iv), the definition is invariant on the choice of the points in the subspace.

**Theorem B.1.42.** *Any projective geometry of dimension greater than two (generated by no less than four points) is isomorphic to a projective geometry over a division ring.*

*Proof.* See the work of Seidenberg (1962). □

**Theorem B.1.43 (weak trichotomy theorem).** *For any strongly minimal* M, *either*

*(0) a pseudo-plane is definable in* M
*or one of the following hold:*
*(i) the geometry of* M *is trivial [i.e. for any* $X \subseteq \hat{M}$, $\mathrm{cl}(X) = (X)$ *in* $\hat{M}$];
*(ii) the geometry of* M *is locally projective.*

*Proof.* Assume no pseudo-plane is definable in M and $c$ is a fixed generic element in $M$.

*Claim 1.* For any $x, y \in M$ and $Z \subseteq M$ finite

$$x \in \mathrm{cl}(y, c, Z) \quad \text{implies} \quad \exists z \in \mathrm{cl}(c, Z) : x \in \mathrm{cl}(y, z, c).$$

We may assume that $Z$ is independent over $c$ and proceed by induction on $\#Z$. For $\#Z = 1$, there is nothing to prove.

Suppose $Z = \{z_1, z_2\} \cup Z'$, $x \in \mathrm{cl}(y, c, Z)$, and $y \notin \mathrm{cl}(c, Z)$. Then by the proposition in $M_{Z'}$, either

(i) some quadruple from $x, y, z_1, z_2, c$ is dependent or
(ii) $\exists z \in M \setminus \mathrm{cl}(c)$

$$\mathrm{cl}(z_1, z_2, c) \cup \mathrm{cl}(x, y, c) \supseteq \mathrm{cl}(z, c).$$

In case (i), only $x \in \mathrm{cl}(y, z_1, z_2)$ is possible, which means in $M$ $x \in \mathrm{cl}(y, z_1, \{z_2 Z'\})$. Because $\#\{z_2, Z'\} < \#Z$, by induction hypothesis there is $z \in \mathrm{cl}(z_1, z_2, Z') : x \in \mathrm{cl}(y, z_1, z)$. If then $\mathrm{cdim}\,(y, z_1, z, c) = 3$, we have $x \in \mathrm{cl}(y, z_1, c)$ or $x \in \mathrm{cl}(y, z, c)$, and we get the desired outcome. Otherwise, there is a point $z_1' \in \mathrm{cl}(z_1, z, c) \setminus (\mathrm{cl}(z_1, z) \cup \mathrm{cl}(c, z_1) \cup \mathrm{cl}(c, z))$. By assuming

$x \notin \mathrm{cl}(y, z)$, we have then that in $\{x, y, z, z_1', c\}$ any four points are independent. Again, using the proposition, there must exist $z' \in M \setminus \mathrm{cl}(c)$ such that

$$\mathrm{cl}(z, z_1', c) \cap \mathrm{cl}(x, y, c) \supseteq \mathrm{cl}(z', c).$$

Clearly $z' \in \mathrm{cl}(c, Z) \cap \mathrm{cl}(x, y, c)$, so $x \in \mathrm{cl}(y, z', c)$, and we are done.

In case (ii), $z \in \mathrm{cl}(c, Z)$ and $x \in \mathrm{cl}_{z'}(y, z, c)$ [i.e. $x \in \mathrm{cl}(y, c\{z, Z'\})$]. By the induction hypothesis, there is $z' \in \mathrm{cl}(c, z, Z')$ such that $x \in \mathrm{cl}(y, z', c)$. The claim is proved.

*Claim 2.* If $\mathrm{cl}(x, y, c) = \mathrm{cl}(x, c) = \mathrm{cl}(y, c)$ for some $x, y$ independent over $c$, then the geometry $\hat{M}$ is degenerate; that is,

$$\mathrm{cl}(x_0, \ldots, x_n) = \mathrm{cl}(x_0) \cup \cdots \cup \mathrm{cl}(x_n)$$

for any $x_0, \ldots, x_n \in M$.

Indeed, under that assumption, $\mathrm{cl}(x_0, x_1, x_2) = \mathrm{cl}(x_0, x_1) \cup \mathrm{cl}(x_0, x_2)$ for any independent triple. We show first that the claim is true for $n = 1$. Assume towards a contradiction $y \in \mathrm{cl}(x_1, x_2) \setminus (\mathrm{cl}(x_1) \cup \mathrm{cl}(x_2))$. Choose $x_0 \notin \mathrm{cl}(x_1, x_2)$. Then

$$y \in \mathrm{cl}(x_0, x_1, x_2) = \mathrm{cl}(x_0, x_1) \cup \mathrm{cl}(x_0, x_2),$$

but if $y \in \mathrm{cl}(x_0, x_i)$ for $i = 1$ or $i = 2$, then $x_0 \in \mathrm{cl}(y, x_i) = \mathrm{cl}(x_1, x_2)$, and there is a contradiction.

Now we proceed by induction on $n$. Suppose $y \in \mathrm{cl}(x_0, \ldots, x_n)$. Then by claim 1 there is $x \in \mathrm{cl}(x_0, \ldots, x_{n-1})$ such that $y \in \mathrm{cl}(x_n, x, x_0)$. From what is proved already, $y \in \mathrm{cl}(x, x_0) \cup \mathrm{cl}(x_n, x_0)$. Hence $y \in \mathrm{cl}(x_0, \ldots, x_{n-1}) \cup \mathrm{cl}(x_0, x_n) = \mathrm{cl}(x_0, \ldots, x_{n-1}) \cup \mathrm{cl}(x_n) \cup \mathrm{cl}(x_n) = \mathrm{cl}(x_0) \cup \cdots \cup \mathrm{cl}(x_n)$.

This finishes the proof of the claim and of the Theorem. $\qquad \square$

It is more common in model-theoretic literature to call a pre-geometry **locally modular** if it is either trivial or locally projective. The negation of the condition, corresponding to the pseudo-plane case of the theorem, is referred to as the non-locally modular case or the **non-linear** case.

**Theorem B.1.44.** *The geometry of a minimal locally finite structure is either trivial or isomorphic to an affine or a projective geometry over a finite field.*

**Scheme of Proof.** First notice that the structure is saturated strongly minimal. Then we can use a result by Doyen and Hubaut, who state that any finite locally projective geometry of dimension $\geq 4$ and equal number of points on all lines is either affine or projective. Thus by the trichotomy theorem, we need to prove only that there is no pseudo-plane in M. It is done by developing a combinatorial-geometric analysis of the pseudo-plane, assuming it exists. The main tool of the analysis is the notion of the 'degree of a line' which is very similar to the degree of an algebraic curve. See our work (Zilber, 1993) for the proof. □

**Corollary B.1.45.** *Any locally finite geometry satisfying the homogeneity assumption (any bijection between bases can be extended to an automorphism) is either trivial or isomorphic to an affine or a projective geometry over a finite field.*

*Proof.* In an appropriate language, such a geometry can be represented as a minimal locally finite structure. □

## B.2 Trichotomy conjecture

### B.2.1 Trichotomy conjecture

As we have observed in Section A.4, the following are basic examples of uncountably categorical structures in a countable language:

1. trivial structures (the language allows equality only);
2. Abelian divisible torsion-free groups; Abelian groups of prime exponent (the language allows $+$, $=$); vector spaces over a (countable) division ring; and
3. algebraically closed fields in language $(+, \cdot, =)$.

Also, any structure definable in one of these is uncountably categorical in the language which witnesses the interpretation.

The structures definable in algebraically closed fields, for example, are effectively objects of algebraic geometry.

As a matter of fact, the main logical problem after answering the question of J. Los was *what properties of* M *make it $\kappa$-categorical for uncountable $\kappa$?*

The answer is now reasonably clear: *The key factors are that we can measure definable sets by a rank-function (dimension) and that the whole construction is highly homogeneous.*

This gave rise to (geometric) stability theory, used to study structures with good dimensional and geometric properties (see the work of Pillay (1996) and Buechler (1996)). When applied to fields, the stability theoretic approach in many respects is very close to algebraic geometry.

Recall that the combinatorial dimension notions (Definition B.1.12) for finite $X \subset M$ in previous examples are as follows:

1a. trivial structures: **size of** $X$;

2a. Abelian divisible torsion-free groups; Abelian groups of prime exponent; vector spaces over a division ring (**linear dimension**, lin.d.$X$, of the linear space spanned by $X$); and

3a. algebraically closed fields: **transcendence degree**, tr.d.$(X)$.

Dually, one can classically define another type of dimension using the initial one:

$$\dim V = \max\{\text{tr.d.}(\bar{x}) \mid \bar{x} \in V\}$$

for $V \subseteq M^n$, an algebraic variety. The latter type of dimension notion in model-theoretic terms is just the Morley rank.

The example of the theory ACF is also a good illustration of the significance of homogeneity of the structures. Indeed, the transcendence degree makes good sense in any field, and there is quite a reasonable dimension theory for algebraic varieties over a field. However, the dimension theory in arbitrary fields fails if we consider it for wider classes of definable subsets (e.g. the images of varieties under algebraic mappings). In algebraically closed fields, any definable subset is a Boolean combination of varieties, by elimination of quantifiers, which eventually is the consequence of the fact that algebraically closed fields are existentially closed in the class of fields (see Section A.3). The latter effectively means high homogeneity, as an existentially closed structure absorbs any amalgam with another member of the class.

One of the achievements of stability theory is the establishment of a hierarchy of types of structures, levels of stability, and finer classification, which roughly speaking correspond to the level of 'analysability' (Shafarevich, 1974).

The next natural question to ask is *whether there are 'very good' stable structures which are not reducible to (1)–(3)?*

The initial hope of the present author was that the following might hold.

**Trichotomy conjecture (Zilber, 1984).**
*The geometry of a strongly minimal structure* M *is either (i) trivial, (ii) locally projective (see Theorem B.1.41), or (iii) isomorphic to a geometry of an algebraically closed field.*

*In particular, if neither (i) nor (ii) is the case, then (iv) there is an algebraically closed field* K *definable in* M *and the only structure induced on* K *from* M *is definable in the field structure itself (the purity of the field).*

The ground for the conjecture was the weak trichotomy theorem (B.1.43) and more importantly the general belief that logically perfect structures could not be overlooked in the natural progression of mathematics. Allowing some philosophical licence here, this was also a belief in a strong logical pre-determination of basic mathematical structures.

Although the trichotomy conjecture proved to be false in general (Hrushovski, 1993, and see also the next sub-section) it turned out to be true in many important classes. The class of *Zariski geometries* is the main class for which this has been proved.

Another situation where the trichotomy principle holds (adapted to the non-stable context) is the class o-minimal structures (Peterzil and Starchenko, 1996).

As was mentioned previously, Hrushovski found a counterexample to the trichotomy conjecture in general. In fact, Hrushovski introduced a new construction which has become a source of a great variety of counterexamples.

### B.2.2 Hrushovski's construction of new stable structures

Suppose we have a class of usually elementary structures $\mathcal{H}$ with a good (combinatorial) dimension notion $d(X)$ for finite subsets of the structures. We want to introduce a new function or relation on $M \in \mathcal{H}$ so that the new structure gets a good dimension notion.

The main principle which produces the desirable effect is that of the free fusion. That is, the new function $f$ should be related to the old $L$-structure in as free of a way as possible. At the same time, we want the structure to be homogeneous. Hrushovski found an effective way of writing down the condition: *the number of explicit dependencies in $X$ in the new structure must not be greater than the size (the cardinality) of $X$.*

The explicit $L$-dependencies on $X$ can be counted as $L$-co-dimension, $\text{size}(X) - d(X)$. The explicit dependencies coming with a new relation or function are the ones given by simplest 'equations', basic formulas.

So, for example, if we want a *new unary function $f$ on a field*, the condition should be

$$\text{tr.d.}(X \cup f(X)) - \text{size}(X) \geq 0, \tag{B.7}$$

because in the set $Y = X \cup f(X)$ the number of explicit field dependencies is $\text{size}(Y) - \text{tr.d.}(Y)$, and the number of explicit dependencies in terms of $f$ is $\text{size}(X)$.

For example, if we want to put a *new ternary relation $R$ on a field*, then the condition would be

$$\text{tr.d.}(X) - r(x) \geq 0, \tag{B.8}$$

where $r(X)$ is the number of triples in $X$ satisfying $R$.

The very first of Hrushovski's examples (1993) introduces just a *new structure of a ternary relation*, which effectively means putting new relation on the trivial structure. Then we have

$$\text{size}(X) - r(X) \geq 0. \tag{B.9}$$

If we introduce an automorphism $\sigma$ on the field (*difference fields*, Chatzidakis and Hrushovski, 1999), then we have to count

$$\text{tr.d.}(X \cup \sigma(X)) - \text{tr.d.}(X) \geq 0, \tag{B.10}$$

and the inequality here always holds, so is not really a restriction in this case.

Similarly, for *differential fields* with the differentiation operator $D$ (Marker, 2000), where again we trivially have

$$\text{tr.d.}(X \cup D(X)) - \text{tr.d.}(X) \geq 0. \tag{B.11}$$

The left-hand side in each of the inequalities (B.7)–(B.11), denoted $\delta(X)$, is a counting function, which is called **pre-dimension**, because it satisfies some of the basic properties of a (combinatorial) dimension notion.

At this point, we have carried out the first step of Hrushovski's construction:

(Dim) we introduced the class $\mathcal{H}_\delta$ of structures with a new function or relation, and the extra condition

(GS)    $\delta(X) \geq 0$    for all finite $X$.

(GS) here stands for 'generalised schanuel', the reason for which is given below. The condition (GS) allows us to introduce another counting function with respect to a given structure $M \in \mathcal{H}_\delta$,

$$\partial_M(X) = \min\{\delta(Y) : X \subseteq Y \subseteq_{\text{fin}} M\}.$$

Now the appropriate notion of embedding in $\mathcal{H}_\delta$ is that of a **strong embedding**, written as $M \leq L$, which means that for every finite $X \subseteq M$, $\partial_M(X) = \partial_L(X)$.

The next step in Hrushovski's construction is marked (EC): Using the inductiveness of the class, construct an existentially closed structure in $(\mathcal{H}_\delta, \leq)$.

If the class has the amalgamation property, then the existentially closed (EC) structures are sufficiently homogeneous. Also *for* M *existentially closed in the class*, $\partial_M(X)$ *becomes a (combinatorial) dimension notion*.

So, if also the sub-class of existentially closed structures is axiomatizable, one can rather easily check that the existentially closed structures are $\omega$-stable. This is the case for examples (B.7)–(B.9) and (B.11).

Another consequence of first-order axiomatisability of the class of EC-structures, by Theorem A.3.5, is model completeness in an expansion of the language where $\leq$ is the same as $\preccurlyeq$. In the original language, this results in the theory of M to have elimination of quantifiers to the level of Boolean combinations of ∃-formulas; equivalently, every definable subset in $M^n$ is a Boolean combination of projective sets.

In more general situations, the class of EC structures may be unstable but still with reasonably good model-theoretic properties.

Notice that though condition (GS) is trivial in examples (B.10)–(B.11), the derived dimension notion $\partial$ is non-trivial. In both examples $\partial(x) > 0$ iff the corresponding rank of $x$ is infinite (which is the SU-rank in algebraically closed difference fields and the Morley rank, in differentially closed fields).

The dimension notion $\partial$ for finite subsets, similar to the example (3a), gives rise to a dual dimension notion for definable subsets $S \subseteq M^n$ over a finite set of parameters $C$:

$$\dim S = \max\{\partial(\{x_1, \ldots, x_n\}/C) : \langle x_1, \ldots, x_n \rangle \in S\}.$$

There is one more stage in Hrushovski construction (called the **collapse**): picking up a sub-structure $M_\mu \subset M$ which approximates M in some special way and has the property of finiteness of rank notion. We are not going to discuss this step of the construction in this book.

The infinite dimensional structures emerging after step (EC) in natural classes we call *natural Hrushovski structures*.

It follows immediately from the construction that the class of natural Hrushovski structures is singled out in $\mathcal{H}$ by three properties: the generalised Schanuel property (GS), the property of existentially closedness (EC), and the property (ID), stating the existence of $n$-dimensional subsets for all $n$.

It takes a bit more model-theoretic analysis, as is done by Hrushovski (1993), to prove that in examples (B.7)–(B.9), and in many others, (GS), (EC), and (ID) form a complete set of first-order axioms.

In 2000, the present author extended the use of Hrushovski's construction to non-elementary (non-first-order) languages in connection to the issues discussed next. It turned out that in this context one more property of structures in question is relevant. This is the **countable closure property**,

$$\text{(CCP)} \quad \dim S = 0 \ \Rightarrow \ \text{card } S(M) \leq \aleph_0.$$

This property is typically definable in the language with the quantifier 'there exist uncountably many $v$ such that ...'.

Once Hrushovski found the counterexamples, the main question that has arisen is whether those seemingly pathological structures demonstrate the failure of the principle in general or if there is a classical context that the counterexamples fit in.

Fortunately, there are good grounds to pursue the latter point of view. We start with one more example of Hrushovski construction.

### B.2.3 Pseudo-exponentiation

We want to put a new function ex on a field $K$ of characteristic zero, so that ex is a homomorphism from the additive into the multiplicative groups of the field:

$$(\text{EXP}) \quad \text{ex}(x_1 + x_2) = \text{ex}(x_1) \cdot \text{ex}(x_2).$$

Then the corresponding pre-dimension on new structures $K_{\text{ex}} = (K, +, \cdot, \text{ex})$ must be

$$\delta(X) = \text{tr.d.}(X \cup \text{ex}(X)) - \text{lin.d.}(X) \geq 0, \quad (\text{GS})$$

where lin.d.$(X)$ is the linear dimension of the $\mathbb{Q}$-span of $X$.

Equivalently, this (GS) can be stated as follows: *assuming that $x_1, \ldots, x_n$ are $\mathbb{Q}$-linearly independent,*

$$\text{tr.d.}(x_1, \ldots, x_n, \text{ex}\, x_1, \ldots, \text{ex}\, x_n) \geq n.$$

This is known in the case $K$ as the field of complex numbers, and ex = exp is the *Schanuel conjecture* (see Lang, 1966).

Start now with the class $\mathcal{H}(\text{ex})$, consisting of structures $K_{\text{ex}}$ satisfying the Schanuel conjecture (GS) and the additional property that the kernel ker = $\{x \in K : \text{ex}(x) = 1\}$ is a cyclic subgroup of the additive group of the field $K$, which we call the **standard kernel**. This class is non-empty and can be described as a sub-class of an elementary class defined by omitting countably many types (Zilber, 2005b).

**Theorem B.2.1.** *In every uncountable cardinality $\kappa$, there is a unique field $K_{\text{ex}}$ with pseudo-exponentiation satisfying the Schanuel condition (GS), existential closedness condition (EC), and the countable closure property (CCP). In other words, $K_{\text{ex}}$ is $\kappa$-categorically axiomatisable by axioms ACF$_0$, EXP, EC, GS, and CCP.*

The theorem is proved using Shelah's stability and categoricity theory for non-elementary classes and some non-trivial arithmetic of fields. Note that the existential closedness condition, typically for structures obtained by Hrushovski's construction, entails the following.

**Theorem B.2.2.** *Two tuples in $K_{\text{ex}}$ are conjugated by an automorphism of the structure iff their projective types coincide.*

By the obvious analogy with the structure $\mathbb{C}_{\text{exp}} = (\mathbb{C}, +, \cdot, \exp)$ on the complex numbers, we conjecture that $\mathbb{C}_{\text{exp}}$ is isomorphic to the unique structure $K_{\text{ex}}$ of cardinality $2^{\aleph_0}$. Note that this is a very ambitious conjecture because it includes Schanuel's conjecture. Moreover, the property (EC) not known for $\mathbb{C}_{exp}$ becomes a new conjecture, essentially stating that any system of exponential-polynomial equations has a solution as long as this fact does not contradict Schanuel's conjecture in a certain direct way. A related result is the main theorem given by Ward Henson and Rubel (1984).

Based on the analysis of pseudo-exponentiation and other similar examples produced by Hrushovski's construction, one starts to hope that though the trichotomy conjecture in full generality is false, some more general classification principle, supposedly referring to *analytic* prototypes (such as the structure $\mathbb{C}_{exp}$), still could be true.

# References

Aschenbrenner, M., Moosa, R., and Scanlon, T. (2006). Strongly minimal groups in the theory of compact complex spaces. *J. Symbolic Logic*, **71**(2), 529–52.

Baldwin, J. T. (1973). $\alpha_T$ is finite for $\aleph_1$-categorical. $T$. *Trans. Amer. Math. Soc.*, **181**, 37–51.

Baldwin, J. (Forthcoming). Categoricity.

Bays, M. and Zilber, B. (2007). *Covers of Multiplicative Groups of Algebraically Closed Fields of Arbitrary Characteristic*. arXiv: math.AC/0401301. (To appear in *J. London Math. Society*.)

Brown, K. and Goodearl, K. (2002). *Lectures on Algebraic Quantum Groups*. Birkhauser.

Bosch, S., Guentzer, U., and Remmert, R. (1984). *Non-Archimedean Analysis*. Volume 261 of *Grundlehren der Mathematischen Wissenschaften*. Berlin: Springer.

Bouscaren, E. (1989). Model theoretic versions of Weil's theorem on pregroups. In *The Model Theory of Groups*. Volume 11 of *Notre Dame Math Lectures*. Notre Dame, IN: University of Notre Dame Press, pp. 177–85.

Buechler, S. (1996). *Essential Stability Theory: Perspectives in Mathematical Logic*. Berlin: Springer.

Chatzidakis, Z. and Hrushovski, E. (1999). Model theory of difference fields. *Trans. Amer. Math. Soc.*, **351**(8), 2997–3071.

Cherlin, G. (2002). Simple groups of finite Morley rank. In *Model Theory and Applications*. Volume 11 of *Quad. Mat.* Rome: Aracne, pp. 41–72.

Danilov, V. I. (1994). Algebraic varieties and schemes. In *Algebraic Geometry I*. Volume 23 of *Encyclopaedia of Mathematical Sciences*.

De Piro, T. (2004). *Zariski Structures and Algebraic Geometry*. ArXiv: math.AG/0402301.

Gavrilovich, M. (2006). *Model Theory of the Universal Covering Spaces of Complex Algebraic Varieties*. PhD thesis, Oxford University, Oxford, UK.

Grauert, H. and Remmert, R. (1984). *Theory of Stein Spaces*. Berlin: Springer-Verlag.

Grossberg, R. (2002). Classification theory for abstract elementary classes. In *Logic and Algebra*. Volume 302 of *Contemp. Math.* Providence, RI: American Mathematical Society, pp. 165–204.

Gunning, R. C. and Rossi, H. (1965). *Analytic Functions of Several Complex Variables.* Princeton Hall Inc., NJ.

Hartshorne, R. (1977). *Algebraic Geometry.* Berlin: Springer Verlag.

Hilbert, D. and Ackermann, W. (1950). *Principles of Theoretical Logic.* Chelsea.

Hodges, W. (1993). *Model Theory.* Cambridge: Cambridge University Press.

Hrushovski, E. (1993). A new strongly minimal set. *Annals of Pure and Applied Logic,* **62,** 147–66.

Hrushovski, E. (1996). The Mordell–Lang conjecture for function fields. *J. Amer. Math. Soc.,* **9,** 667–90.

Hrushovski, E. and Zilber, B. (1993). Zariski geometries. *Bull. AMS,* **28,** 315–23.

Kirby, J. (2008). *The Theory of the Exponential Differential Equations of Semi-Abelian Varieties.* arXiv:0708.1352.

Kueker, D. (1975). Back-and-forth arguments and infinitary logics. In *Infinitary Logic: In Memoriam Carol Karp.* Volume 492 of *Lecture Notes in Math.* Berlin: Springer, pp. 17–71.

Lang, S. (1966). *Introduction to Transcendental Numbers.* Reading, MA: Addison-Wesley.

Marker, D. (2000). Model theory of differential fields. In *Model Theory, Algebra, and Geometry.* Volume 39 of *Math. Sci. Res. Inst. Publ.* Cambridge: Cambridge University Press, pp. 53–63.

Marker, D. (2002). *Model Theory: An Introduction.* Volume 217 of *Graduate Texts in Mathematics.* New York: Springer.

Moosa, R. (2005a). The model theory of compact complex spaces. In *Logic Colloquium 2001.* Volume 20 of *Lect. Notes Log.* Urbana, IL: Association Symbol. Logic, pp. 317–49.

Moosa, R. (2005b). On saturation and the model theory of compact Kähler manifolds. *J. Reine Angew. Math.,* **586,** 1–20.

Mycielski, J. 1964. Some compactification of general algebras. *Colloq. Math.,* **13,** 1–9.

Peatfield, N. and Zilber, B. (2004). Analytic Zariski structures and the Hrushovski construction. *Annals of Pure and Applied Logic,* **132,** 127–80.

Peterzil, Y. and Starchenko, S. (1996). Geometry, calculus, and Zilber's conjecture. *Bull. Symbolic Logic,* **2**(1), 72–83.

Pillay, A. (1996). *Geometric Stability Theory.* Volume 32 of *Oxford Logic Guides.* New York: Clarendon Press.

Pillay, A. (2002). Two remarks on differential fields. In *Model Theory and Applications.* Volume 11 of *Quad. Mat.* Rome: Aracne, pp. 325–47.

Pillay, A. and Scanlon, T. (2002). Compact complex manifolds with the DOP and other properties. *J. Symbolic Logic,* **67**(2), 737–43.

Pillay, A. and Scanlon, T. (2003). Meromorphic groups. *Trans. Amer. Math. Soc.,* **355**(10), 3843–59.

Poizat, B. (1987). *Groups Stables.* Paris: Nur al-Mantiq Wal-Ma'rifah.

Poizat, B. (2001). L'égalité au cube. *J. Symbolic Logic,* **66**(4), 1647–76.

Robinson, A. (1977). *Complete Theories,* 2nd ed. Amsterdam: North-Holland Publishing Co.

Shafarevich, I. R. (1977). *Basic Algebraic Geometry,* study ed. Translated from the Russian by K. A. Hirsch. Berlin: Springer-Verlag.

Seidenberg, A. (1962). *Lectures in Projective Geometry.* D. van Nostrand Co.

Smith, L. (2008). *Toric Varieties as Analytic Zariski Structures*. PhD thesis, Oxford University, Oxford, UK.

Ward Henson, C. and Rubel, L. A. (1984). Some applications of Nevanlinna theory to mathematical logic: Identities of exponential functions. *Trans. Amer. Math. Soc.*, **282**(1), 1–32.

Weglorz, B. (1966). Equationally compact algebras I. *Fund. Math.*, **59**, 289–98.

Weil, A. (1962). *Foundations of Algebraic Geometry*. RI: AMS.

Zilber, B. (1974). The transcendence rank of the formulae of an $\aleph_1$-categorical theory (in Russian). *Mat. Zametki*, **15**, 321–9.

Zilber, B. (1984). *The Structure of Models of Uncountably Categorical Theories*, Volume 1, pp. 359–68. Amsterdam: PWN–North-Holland.

Zilber, B. (1993). *Uncountably Categorical Theories*. Volume 117 of *AMS Translations of Mathematical Monographs*. Providence, RI: AMS.

Zilber, B. (1997). Generalized analytic sets. *Algebra i Logika*, **36**(4), 361–80.

Zilber, B. (2002a). Exponential sums equations and the Schanuel conjecture. *J. London Math. Soc. 2*, **65**(1), 27–44.

Zilber, B. (2002b). Model theory, geometry, and arithmetic of the universal cover of a semi-Abelian variety. In *Model Theory and Applications*, ed. L. Belair et al. Volume 11 of *Quaderni di Matematica*, pp. 427–58. Caserta.

Zilber, B. (2003). Raising to powers in algebraically closed fields. *J. Math. Log.*, **3**(2), 217–38.

Zilber, B. (2004a). Bi-coloured fields on the complex numbers. *J. Symbolic Logic*, **69**(4), 1171–86.

Zilber, B. (2004b). *Raising to Powers Revisited*. author's web-page.

Zilber, B. (2005a). A categoricity theorem for quasi-minimal excellent classes. In *Logic and Its Applications*. Volume 380 of *Contemp. Math.*, pp. 297–306. Providence, RI: Amer. Math. Soc.

Zilber, B. (2005b). Pseudo-exponentiation on algebraically closed fields of characteristic zero. *Ann. Pure Appl. Logic*, **132**(1), 67–95.

Zilber, B. (2006). Covers of the multiplicative group of an algebraically closed field of characteristic zero. *J. London Math. Soc. 2*, **74**, 41–58.

Zilber, B. (2008a). A class of quantum Zariski geometries. In *Model Theory with Applications to Algebra and Analysis, I*, ed. Z. Chatzidakis, H. D. Macpherson, A. Pillay, and A. J. Wilkie. Volume 349 of *LMS Lecture Notes Series*. Cambridge: Cambridge University Press.

Zilber, B. (2008b). *Analytic Zariski Structures, Predimensions, and Non-elementary Stability*. MODNETpreprints-132.

# Index

Printed in the United States
By Bookmasters